Macromolecular Symposia

Symposium Editor: T. Kajiyama

Editor: I. Meisel
Associate Editor: S. Spiegel
Assistant Editors: H. Beattie, C.S. Kniep

Executive Advisory Board: M. Antonietti, M. Ballauff, H. Höcker,
K. Kremer, H.E.H. Meijer, R. Mülhaupt,
A.D. Schlüter, H.W. Spiess, G. Wegner

159

pp. 1–266

October 2000

Macromolecular Symposia publishes lectures given at international symposia and is issued irregularly, with normally 14 volumes published per year. For each symposium volume, an Editor is appointed. The articles are peer-reviewed. The journal is produced by photo-offset lithography directly from the authors' typescripts.
Further information for authors can be obtained from:
Editorial office "Macromolecular Symposia"
Wiley-VCH
P. O. Box 10 11 61, D-69451 Weinheim,
Germany
or for parcel and courier services: Pappelallee 3, D-69469 Weinheim,
Germany
Tel. +49 (0) 62 01/6 06-2 34 or -2 38; Fax +49 (0) 62 01/6 06-3 09 or 5 10; macromol@wiley-vch.de
http://www.wiley-vch.de/home/macrosymp
Suggestions or proposals for conferences or symposia to be covered in this series should also be sent to the Editorial office at the address above.

Macromolecular Symposia:
Annual subscription rates 2001
Germany, Austria € 1098 (DM 2147,50); Switzerland SFr 1808; other Europe € 1098; outside Europe US $ 1328.
Macromolecular Package, including Macromolecular Chemistry & Physics (18 issues), Macromolecular Rapid Communications (18 issues), Macromolecular Theory & Simulations (9 issues) is also available. Details on request.
Packages including Macromolecular Symposia and Macromolecular Materials & Engineering are also available. Details on request.
Single issues and back copies are available. Please, inquire for prices.

Orders may be placed through your bookseller or directly at the publishers: WILEY-VCH Verlag GmbH, P. O. Box 10 11 61, D-69451 Weinheim, Germany, Tel.: (0 62 01) 6 06–0, Telefax (0 62 01) 60 61 17, Telex 46 55 16 vchwh d. E-mail: subservice@wiley-vch.de

Macromolecular Symposia (ISSN 1022-1360) is published with 14 volumes per year by WILEY-VCH Verlag GmbH, P. O. Box 10 11 61, D-69451 Weinheim, Germany. Air freight and mailing in the USA by Publications Expediting Inc., 200 Meacham Ave., Elmont, NY 11003. Periodicals postage pending at Jamaica, NY 11431. POSTMASTER: send address changes to Macromolecular Symposia, Publications Expediting Inc., 200 Meacham Ave., Elmont, NY 11003.

© WILEY-VCH Verlag GmbH, D-69469 Weinheim, Germany, 2000
Printing: Strauss Offsetdruck, Mörlenbach. Binding: Wilh. Osswald, Neustadt

The Society of Polymer Science, Japan

Selected invited lectures presented at the

7th SPSJ International Polymer Conference (IPC 99)

"Polymer Science and Industrial Research in the Fast-changing Age"

held in Yokohama, Japan
October 26–29, 1999

Symposium Editor

T. Kajiyama
Graduate School of Engineering, Kyushu University
Hakozaki, Higashi-ku, Fukuoka 812-8581, Japan

Copyright © 2000 WILEY-VCH Verlag GmbH
ISBN 3-527-30138-0

Contents of Macromol. Symp. 159

**7th SPSJ International Polymer Conference (IPC 99)
Yokohama, Japan, 1999**

Preface
Tisato Kajiyama

* The asterisk indicates the name of the author to whom inquiries should be addressed

Author Index

Preface

This special issue contains most of the invited lectures presented at the 7th SPSJ International Polymer Conference (IPC 99). It was held at Yokohama Prince Hotel during October 26–29, 1999. The Theme of this conference was "Polymer Science and Industrial Research in the Fast-changing Age", and the numbers of participants and presentations were 579 from 24 countries and 189, respectively. The conference was composed of the three parts of Industrial Research, Selected and General Topics as follows.

(1) Industrial Research-Strategy for Polymer R&D and its KFS (Key for Success)
(2) Selected Topics
   *New Concept in Polymerization
   *Surface and Interface Engineering of Polymeric Materials
   *Gels and Network Polymers
   *Polymer Materials for Batteries
   *Polymers and Environment
(3) General Topics
   *Polymer Synthesis and Reactions
   *Structure and Physical Properties of Polymers
   *Polymer Processing
   *Functional Polymers
   *High Performance Polymers
   *Bio-Related Polymers

As mentioned above, a wide range of polymer science as well as industrial research was covered to understand the present situation of polymer research and to search the future interests and trends on polymer science and industry. IPC 99 contributed to strength academic ties among world-wide polymer research groups on a scientific and personal level, through enthusiastic discussions in the conference room and the hotel lobby.

The Organizing and Program Committees would like to express acknowledgement for a great success of IPC and important scientific contribution to all participants and contributors.

Tisato Kajiyama
Chairman
Program Committee
IPC 99

Novel Ring-Opening Polymerization and Its Application to Polymeric Materials

Takeshi Endo,* Fumio Sanda [†]

*Department of Polymer Science and Engineering, Faculty of Engineering, Yamagata University, 4-3-16 Jonan, Yonezawa, Yamagata 992-8510, Japan

[†] Chemical Resources Laboratory, Tokyo Institute of Technology

4259 Nagatsuta-cho, Midori-ku, Yokohama 226-8503, Japan

SUMMARY: Ring-opening polymerization of cyclic carbonates was studied. Volume expansion during polymerization of cyclic carbonates could be explained by the difference between intermolecular interaction of monomers and polymers, which was evaluated by dipole moment. A seven-membered cyclic carbonate polymerized much faster than a six-membered one. Bulky substituents of six-membered cyclic carbonates increased the equilibrium monomer concentration. A novel initiator system, combination of alcohol and trifluoroacetic acid, efficiently polymerized cyclic carbonates.

Introduction

Chemistry of ring-opening polymerization was established in 1950s. The history of ring-opening polymerization is relatively shorter compared with vinyl polymerization and polycondensation, which have been widely developed both from scientific and industrial points of view.[1] Nowadays, ring-opening polymerization plays an important role in industry such as production of nylon-6. Ring-opening polymerization can introduce functional groups like ether, ester, amide, and carbonate into the polymer main chain, which cannot be attained by vinyl polymerization affording polymers only with C-C main chain. Polymers obtained by ring-opening polymerization can be also prepared by polycondensation in most cases, but following control is possible in ring-opening polymerization, which is difficult in polycondensation.

- Equivalency of two functional groups is important to obtain a high molecular weight polymer in polycondensation. This is automatically achieved in ring-opening polymerization.
- Living polymerization can be expected in ring-opening polymerization. Molecular weight and polydispersity ratio can be controlled.
- Copolymerization with controlled unit sequence (block polymerization) is possible.

- Control of stereo regularity is possible.

Recently, development of novel monomers and catalysts has enabled us to control molecular weights, structure, and configuration of the polymers more precisely than ever before. Cyclic carbonates undergo both cationic and anionic polymerizations to afford the corresponding polycarbonates (Scheme 1),[2] which are expected as biocompatible and bio-degradable polymers.[3] Recently, ultra high molecular weight bisphenol A polycarbonate (> Mw 2,000,000) has been synthesized by the ring-opening polymerization a large-membered bisphenol A-based cyclic carbonate.[4] The obtained polymer is expected as engineering plastics with high impact resistance and thermal stability. As described above, polymerization of cyclic carbonates attracts much attention both from polymerization behavior and function. This article deals with controlled ring-opening polymerization of cyclic carbonates.

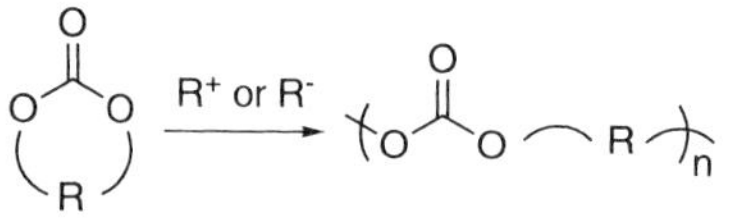

Scheme 1

Volume expansion during ring-opening polymerization of cyclic carbonates

Polymerization of vinyl monomers such as styrene and methyl methacrylate, and cyclic monomers such as epoxides is accompanied by volume shrinkage as large as ~ 20%. The reason can be readily understood from the decrease of intermolecular distance of monomer molecules located at a van der Waals distance to a covalent distance by polymerization. Occurrence of volume shrinkage exhibits serious problems in many industrial applications of polymeric materials, and thereby development of monomers and materials showing volume expansion or zero-shrinkage is one of the most essential subjects in materials science.[2] Several attempts have been made to reduce volume shrinkage like improvement of adhesion method and addition of fillers. However, the problems of lowering of material properties caused by curing shrinkage have not been completely solved yet. We have previously reported that some bicyclic and spiro cyclic monomers such as bicyclo orthoesters, spiro orthoesters, and spiro orthocarbonates undergo cationic double ring-opening polymerization accompanying volume expansion (Scheme 2). Copolymerization of these expandable monomers with thermosetting resins is effective to reduce volume shrinkage during curing process and enhance the product property. These monomers exhibit volume expansion according to principle as follows.   Free volume decreases by polymerization, because the distance between monomer molecules becomes close from van der Waals distance to covalent distance.   Meanwhile, the covalent distance of the monomer transforms into near van der Waals distance by double ring-opening polymerization, which compensates the free volume decrease by polymerization.

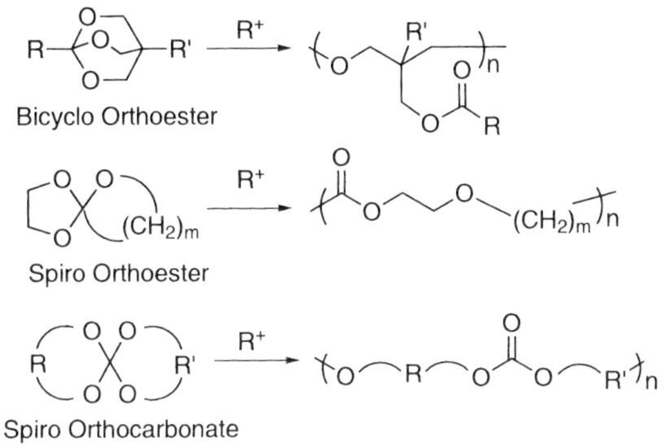

Bicyclo Orthoester

Spiro Orthoester

Spiro Orthocarbonate

Scheme 2

Recently, we have found that cyclic carbonates show volume expansion on polymerization.[2] This is very interesting and significant, because they only undergo single ring-opening polymerization, and therefore cannot be explained by the concept applicable to the expandable bicyclic and spirocyclic monomers described above.   We have noticed that change in intermolecular interaction before and after polymerization can explain the volume expansion of cyclic carbonates, i. e., density decrease by decrease in intermolecular interaction during polymerization is more than density increase resulted by conversion of monomer molecules to the polymer (Scheme 3).   It can be demonstrated that difference in dipole moment between monomer and polymer is available to evaluate the intermolecular interaction.

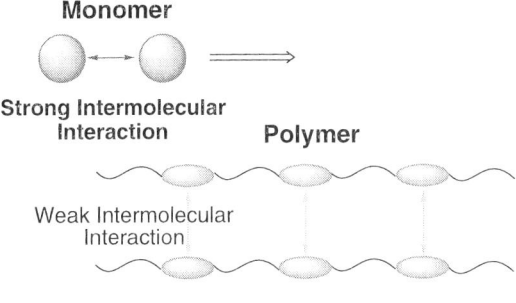

Monomer

Strong Intermolecular Interaction

Polymer

Weak Intermolecular Interaction

Scheme 3

4

We calculated the dipole moment of the monomer and the polymer model, dimethyl carbonate. The cyclic carbonate showed a large value (5.4 debye), while dimethyl carbonate showed a small value (1.0 debye) (Scheme 4). It was suggested that cyclic carbonates show volume expansion due to the large intermolecular interaction by the large dipole moment, and small interaction by the small dipole moment of the polymer.[5] This mechanism of volume expansion of cyclic carbonates is completely different from that of bicyclic and spiro cyclic monomers.

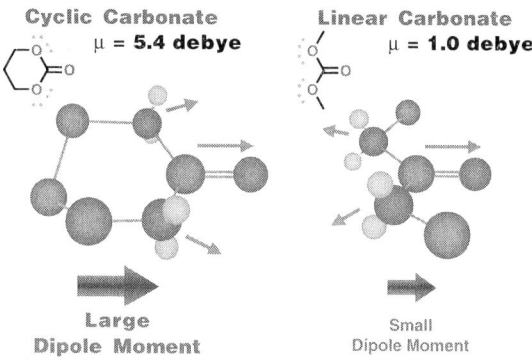

Scheme 4

Cationic and anionic ring-opening polymerizations of seven-membered cyclic carbonates

Many researches on ring-opening polymerization of six-membered cyclic carbonates have been reported concerning substituent effect, control of molecular weight and tacticity, depolymerization, and copolymerization with other monomers. On the contrary, only a few reports of ring-opening polymerization of cyclic carbonates have been reported so far. We examined the ring-opening polymerization of a seven-membered cyclic carbonate to find that the rates of anionic and cationic polymerizations of the seven-membered one were 36 and 100 times faster compared with those of a six-membered one, respectively.[6,7] We further studied the anionic and cationic ring-opening polymerizations of a seven-membered cyclic carbonate

(Ph7CC) substituted with a phenyl group.[8] The anionic polymerization of Ph7CC was carried out with *sec*-BuLi or *tert*-BuOK (1 mol %) as an initiator in tetrahydrofuran (0.8 M) at 20 °C for 1 h to afford the corresponding polycarbonate with M_n of 11,700 ~ 13,400 in 79 ~ 88% yield (Scheme 5). The cationic polymerization of Ph7CC was carried out with TfOMe, TfOEt, or $BF_3 \cdot OEt_2$ as an initiator (1 mol %) in CH_2Cl_2 (1 M) at 20 °C for 4 h to obtain the corresponding polycarbonate with M_n of 3,000-5,600 in 35-63% yield (Scheme 5). The ^{13}C-NMR spectrum of the polymer obtained by anionic polymerization showed three carbonyl carbon signals, while that by cationic polymerization showed one signal, suggesting the selective cleavage of the benzyl carbon-ether oxygen bond due to formation of stable benzyl cation.

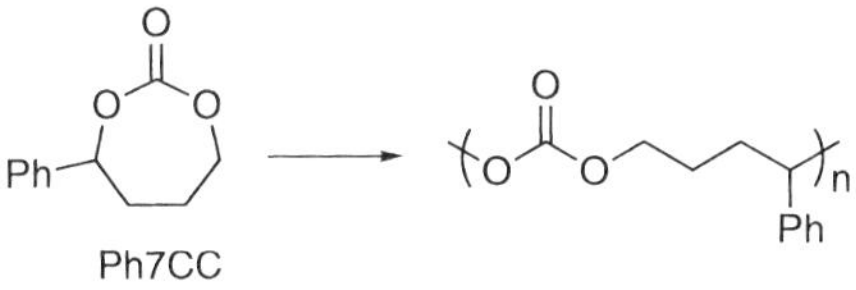

Ph7CC

Scheme 5

Anionic equilibrium polymerization of six-membered cyclic carbonates

Substituent effect on anionic equilibrium polymerization was examined. A cyclic carbonate substituted with aromatic groups showed a higher equilibrium monomer concentration, i. e., low polymerizability. Calculated thermodynamic parameters agreed well with the polymerizability (Scheme 6, Table 1).[9] Introduction of aromatic substituents seemed to distort the carbonate moiety of the polymer main chain, resulting in destabilization of the polymer. The monomer with two phenyl groups at the 5-position did not polymerize at all. It was suggested from the molecular orbital calculation that back-biting reaction was predominate compared with propagation reaction.[10] The copolymerization of this monomer with dimethyl substituted monomer could afford the copolymer.

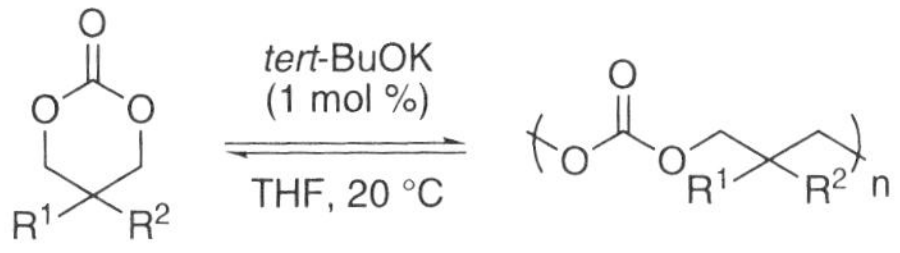

Scheme 6

Table 1.   Equilibrium monomer concentration and thermodynamic parameters in the anionic polymerization of six-membered cyclic carbonates [a]

| monomer | | equilibrium monomer | | | |
R^1	R^2	conc (M)	ΔH_p (kcal/mol)	ΔS_p (cal/K•mol)	ΔG_p (kcal/mol)
H	H	0.02 [b]	-6.3	-10.7	-3.1
Me	Me	0.03	-5.1	-10.5	-2.0
Et	Et	0.08	-4.4	-10.9	-1.2
Me	Ph	0.11	-4.0	-9.7	-1.2
Et	Ph	0.41	-1.2	-2.4	-0.5
Ph	Ph	— [c]	— [c]	— [c]	— [c]

a) Conditions: initiator *tert*-BuOK (1 mol %), solvent THF (initial monomer conc. 0.6 M), temperature 20 °C.   b) Initial monomer conc. 0.05 M.   c) No polymerization.

Ring-opening polymerization of cyclic carbonates with alcohol-acid catalyst

Cyclic carbonates react with several nucleophiles to afford the corresponding ring-opened adducts.   Reaction of cyclic carbonates with amines has been extensively studied, while that with alcohols has not.   We examined the ring-opening polymerization of cyclic carbonates based on the reaction with alcohols in the presence of acid catalysts.   The polymerization of six- and seven-membered cyclic carbonates was carried out with benzyl alcohol or *n*-butanol (1-5 mol %) in the presence of trifluoroacetic acid (1 mol %) in CH_2Cl_2 or toluene (2 M) at 0 or 50 °C.   The polymers with M_n of 2,500-6,800 (M_w/M_n 1.16-1.24) were obtained in 75-93% yield (Scheme 7).[11]   The molecular weight of the polymer increased as the amount of the used alcohol decreased.   No reaction of cyclic carbonates took place with alcohols in the absence of catalysts, but it efficiently proceeded in the presence of acids.   It was suggested that the acids activated the cyclic carbonates by coordination.   For confirmation, we examined the change of chemical shifts of the cyclic carbonates by the addition of trifluoroacetic acid by ^1H and ^{13}C NMR spectroscopy to find the lower field shift of the α and β-methylene protons, and also carbonyl carbon signal.

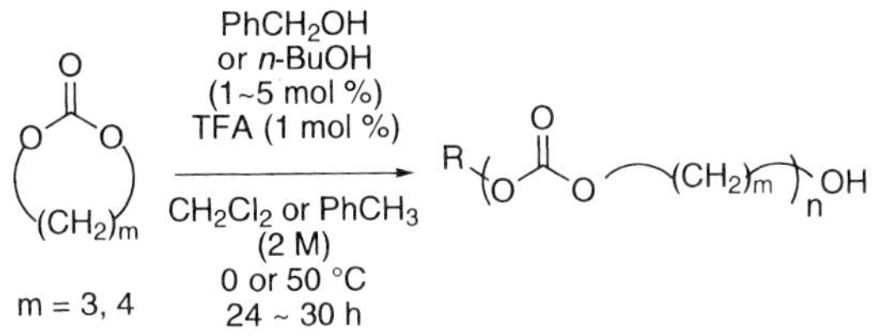

PhCH$_2$OH
or n-BuOH
(1~5 mol %)
TFA (1 mol %)

$\xrightarrow{\quad}$

CH$_2$Cl$_2$ or PhCH$_3$
(2 M)
0 or 50 °C
24 ~ 30 h

m = 3, 4

R$-$O$-$C($=$O)$-$O$-$(CH$_2$)$_m$$-$OH $)_n$

Scheme 7

Conclusion

In this article, we reviewed our recent study on controlled ring-opening polymerization of cyclic carbonates. Further development of novel monomers and initiators is expected, based on the previous information and molecular design utilizing computational chemistry.

References

1. K. J. Ivin, T. Saegusa, *Ring-Opening Polymerization,* Elsevier Applied Science Publishers, London 1984
2. T. Takata, T. Endo, Ionic Polymerization of Oxygen-Containing Bicyclic, Spirocyclic, and Related Expandable Monomers in *Expanding Monomers: Synthesis, Characterization and Applications* R. K. Sadhir, R. M. Luck, Eds. CRC Press, Boca Raton 1992, p. 63
3. S. Matsumura, K. Tsukada, K. Toshima, *Macromolecules* **30**, 3122 (1997)
4. J. Sugiyama, R. Nagahata, M. Goyal, M. Asai, M. Ueda, K. Takeuchi, *ACS Polym. Prepr.* **40(1)**, 90 (1998)
5. T. Takata, F. Sanda, T. Ariga, H. Nemoto, T. Endo, *Macromol. Rapid Commun.* **18**, 461 (1997)
6. J. Matsuo, F. Sanda, T. Endo, *J. Polym. Sci., Part A: Polym. Chem.* **35**, 1375 (1997)
7. J. Matsuo, F. Sanda, T. Endo, *Macromol. Chem. Phys.* **199**, 97 (1998)
8. J. Matsuo, F. Sanda, and T. Endo, *Macromol. Chem. Phys.* **201**, 585 (2000)
9. J. Matsuo, K. Aoki, F. Sanda, T. Endo, *Macromolecules* **31**, 4432 (1998)
10. J. Matsuo, F. Sanda, T. Endo, *Macromol. Chem. Phys.* **199**, 2489 (1998)
11. J. Matsuo, S. Nakano, F. Sanda, T. Endo, *J. Polym. Sci., Part A: Polym. Chem.* **36**, 2463 (1998)

Macromol. Symp. **159**, 9–17 (2000)

Progress of Olefin Polymerization by Metallocene Catalysts

Walter Kaminsky*, Volker Scholz, Ralf Werner

Institute for Technical and Macromolecular Chemistry, University of Hamburg, Bundesstr. 45, 20146 Hamburg, Germany

SUMMARY: New C_1-symmetric metallocenes such as [Me$_2$C(PhCp)(Flu)ZrCl$_2$, [Me$_3$Pen(Flu)]ZrCl$_2$, [PhMe$_3$Pen(Flu)]ZrCl$_2$ were synthesized and used for the polymerization of propene by higher polymerization temperatures. Different polypropylene micro structures were obtained. Important for industrial processes are the high molecular weights of the polymers produced by the pentalenelike catalysts, which are very stable by higher temperatures. For synthesis of syndiotactic polystyrene and new substituted half-sandwich titanocenes are used such as 1,3-Me$_2$-CpTiCl$_3$, Me$_4$CpTiCl$_3$, PhCpTiCl$_3$, cyclohexyl-CpTiCl$_3$. If they are fluorinated, the activity for the production of syndiotactic polystyrene can be increased 10 times. The synthesized polymer shows a high melting point of 275 °C.

Introduction

Metallocene and half sandwich complexes are in combination with methylaluminoxane (MAO) active catalysts for the production of precisely designed polyolefins and engineering plastis [1,3]. Especially C_s or C_1 symmetric zirconocenes with an isopropylidene bridge and substituted cyclopentadienyl and fluorenyl ligands are of great interest as catalysts for the polymerization of propene, which leads to syndiotactic and hemiisotactic polymers [4-7]. They can be synthesized without achiral side products. They are also very active for the polymerization of ethene and copolymers by ethene and other 1-olefins (Tab. 1).

Table 1 Ethene Homopolymerization by Different [Me$_2$C(3-RCp)(Flu)]ZrCl$_2$/MAO Catalysts at 30 °C in Toluene; [Zr] = 5 x 10^{-6} mol/l, Al:Zr = 7800

	Activity (kg PE/mol Zr x h)		
[Ethene]	R = Me	iPr	tBu
0,118	640	640	21 000
0,236	1 700	1 400	87 000
0,472	2 900	6 600	124 000
0,708	4 700	9 500	210 000

CCC 1022-1360/00/$ 17.50+.50/0

Only by changing the substitution at the Cp-ring, the activity can be increased by the factor of 20 from i-propyl to t-butyl. For an industrial use some disadvantages have to be solved. The polymers often show a low molecular mass, and the catalysts are not very stable at higher temperatures and deactivate fast. In other cases these catalysts are not active enough and give only low stereoselectivities.

Therefore there is an interest to increase the activity and to make the metallocenes more stable for higher temperatures [7]. It is known from the bisindenylzirconocenes that substitution by phenyl groups increases the activity [8]. A stronger bridge should increase the thermal stability. X-ray structures and calculations showed that the angle between the rings of the zirconocene can be influenced (Fig. 1).

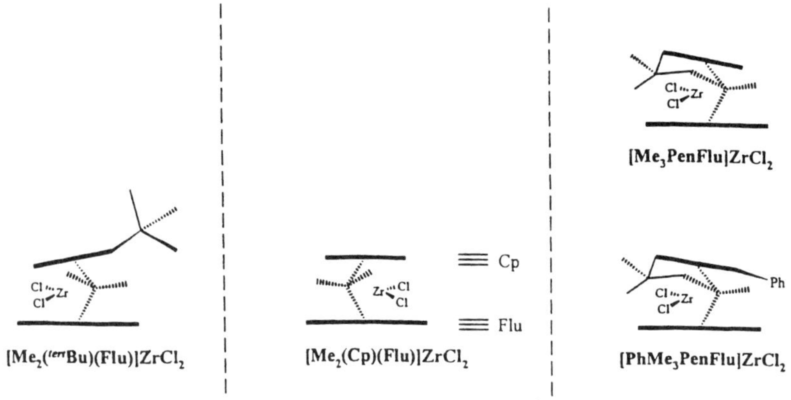

Fig. 1 Angle between the rings of different C_s and C_1 symmetric metallocenes

The pentalenelike compounds give more space for the insertion of longer chained 1-olefins, or cyclic olefins and should be useful for copolymerization.

Substitution of the cyclopentadienyl ring in [Me$_2$C(tert-Bu Cp)(Flu)]ZrCl$_2$ yields ethene/norbornene copolymers with an alternating structure, because the rigid norbornene can only be inserted from the open side of the metallocene. By variation of the polymerization parameters, copolymers with glass transition temperatures above 180 °C and molecular weights >100 000 are synthesized. Other substitutions lead to block micro-structures of the copolymers.

Results and Discussions

Propene Polymerization

Four new ansa zirconocenes (1-4) were synthesized to increase the activity and the thermal stability of the metallocenes for the polymerization of ethene and propene (Fig. 2)

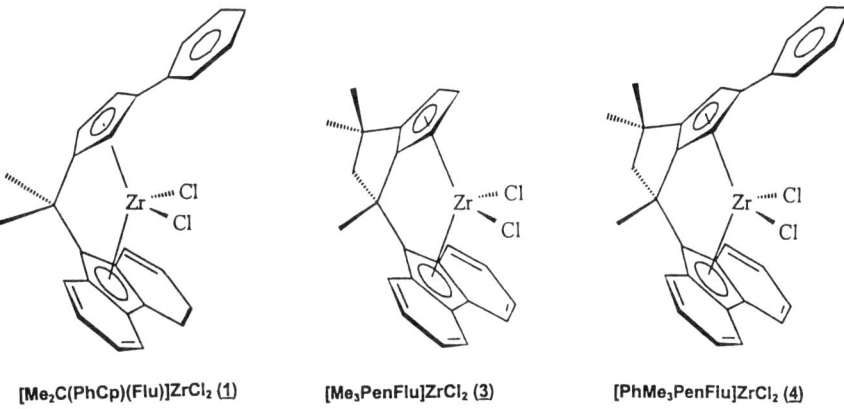

[Me₂C(PhCp)(Flu)]ZrCl₂ (1)   **[Me₃PenFlu]ZrCl₂ (3)**   **[PhMe₃PenFlu]ZrCl₂ (4)**

Fig. 2 Structures of C_1-symmetric zirconocenes

The compounds [1-(Fluorenyl)-1,3,3-trimethyl-tetrahydropentalenyl]zirconium dichloride ([Me₃PenFlu]ZrCl₂, **3**) and [1-(Fluorenyl)-1,3,3-trimethyl-5-phenyl-tetrahydropentalenyl] zirconium dichloride ([PhMe₃PenFlu]ZrCl₂, **4**) show a more stereorigid bridge than the zorconocenes 1 and 2, which have an isopropyl bridge.

The zirconocenes activated by MAO are used for the polymerization of propene under the same conditions and compared with some other known complexes (Tab. 2).

The characteristics of the polymerization of propene with the four new catalysts were investigated with reference to the influence of temperature and monomer concentration. The significance of these effects was rated by means of design of experiments. Particular reference was paid to the effects on the activity of the catalysts and on the molecular weight and microstructure of the polymers prepared.

Table 2 Results of Polymerizations under Standardized Conditions

Catalyst	Activity[a]	M_v[b]	T_g[c]	T_m[d]	n_{iso}[e]	n_{syn}[f]
1 [Me₂C(PhCp)(Flu)]ZrCl₂	7 500	60	-2.7	-	3.1	4.9
2 [Me₂C(cHCp)(Flu)]ZrCl₂	2 900	70	-2.9	-	3.2	4.8
3 [Me₃PenFlu]ZrCl₂	4 000	480	8.9	133.8	1.5	14.9
4 [PhMe₃PenFlu]ZrCl₂	790	260	-4.0	82,0	9.6	2.6
for comparison:						
5 [Me₂C(tertBuCp)(Flu)]ZrCl₂	1 100	50	n.d.	133.5	n.d.	n.d.
6 [Me₂C(Cp)(Flu)]ZrCl₂	1 600	160	5.1	143.1	1.6	46.0
7 rac-[En(Ind)₂]ZrCl₂	1 700	30	n.d.	136.5	48.5	1.5
8 rac-[Me₂Si(Ind)₂]ZrCl₂	1 900	80	n.d.	148.9	99.0	1.0

Conditions: Temperature: 30 °C; monomer concentration: 1.31 mol/l (propene); volume: 200 ml toluene; cocatalyst: MAO; [a] [kgPP/(molZr · h · $c_{propene}$)]; [b] average viscosity molecular weight [kg/mol]; [c] glass transition temperature [°C]; [d] melting point [°C]; [e] average isotactic chain length; [f] average syndiotactic chain length; n.d.: not detected

The phenyl substituted **1** revealed an activity of up to nine times higher than the rest of the systems to which it was compared. Whereas metallocenes **1** and **2** show a fast thermal deactivation at higher polymerization temperatures, an increase in activity of complexes **3** and **4** was found by raising the temperature; this may be an effect of the very rigid structure of these compounds.

One of the most important results of the work presented is the high molecularweight of the polymers produced by the pentalenelike catalysts **3** and **4**: That are up to nine times the value the isopropylidene bridged zirconocenes reach.

The microstructure of the polymers prepared at 30 °C by catalysts **1** and **2** is, in general, the same (rrrr = 33 %). At high temperatures or low monomer concentrations, however, the product polymerized by metallocene **2** is isotactic. The steric hindrance can explain this unusual phenomenon. While complex **3** produces polymers with a syndiotactic micostructure, metallocene **4** yields an isotactic product. Another remarkable result of this work is the lack of effects of variation in temperature and monomer concentration on the polymer structure produced by catalyst **4** [7].

Of great interest is the activity by higher polymerization temperatures. Fig. 3 compares the activity of the complexes 1 - 4 for the polymerization of propene by different temperatures.

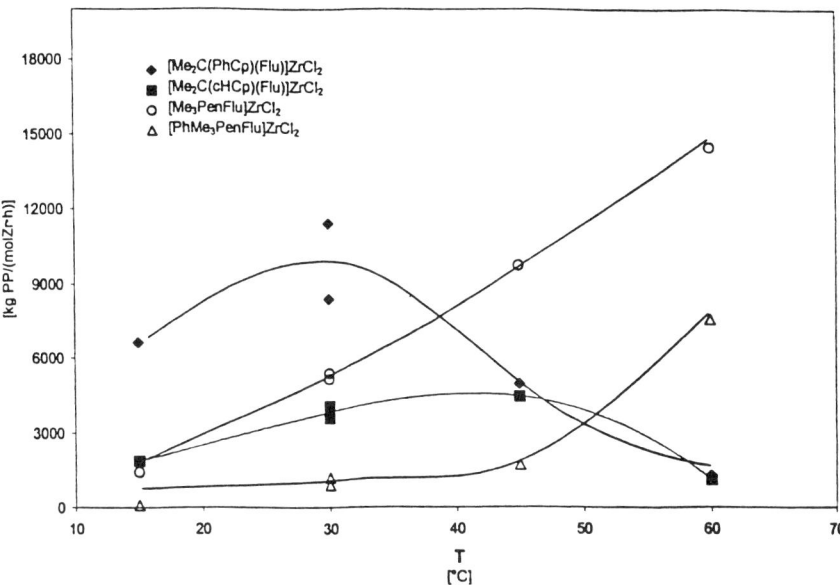

Fig. 3 Polymerization activity of propene in dependence of temperature for different zirconocenes

It can be seen that by low polymerization temperatures (30 °C) the activity of zirconocene 1 with a phenyl substitution at the Cp-ring shows a maximum but by increasing the temperature (60 °C) the pentalene like complex 3 is much active and produces the polypropylene with the higher molecular weight.

Syndiotactic polystyrene

As found by Ishihara, syndiotactic polystyrene can be synthesized by titanium/MAO catalysts [9]. Half sandwich complexes such as $CpTiCl_3$, $Cp*TiCl_3$, or tetra-alkoxy complexes in combination with MAO polymerize styrene affording a highly syndiotactic polymer with a melting point of 275 °C in contrast to the commercial atactic polystyrene with a glass transition point of 100 °C [10-13].

The activity however is much less compared with olefin polymerization so that catalyst costs become important for industrial production. Therefore, it is necessary to enhance the activity of the catalyst.

New substituted half-sandwich titanocenes (1,3-Me$_2$-CpTiCl$_3$, Me$_4$CpTiCl$_3$, PhCpTiCl$_3$, C$_6$H$_{11}$CpTiCl$_3$) were synthesized and used for comparison with other substituted compounds for the polymerization of styrene (Fig. 4)

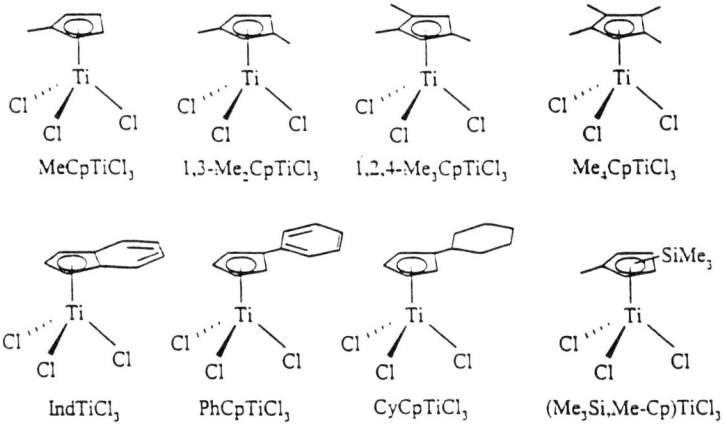

MeCpTiCl$_3$ 1,3-Me$_2$CpTiCl$_3$ 1,2,4-Me$_3$CpTiCl$_3$ Me$_4$CpTiCl$_3$

IndTiCl$_3$ PhCpTiCl$_3$ CyCpTiCl$_3$ (Me$_3$Si,Me-Cp)TiCl$_3$

Fig. 4 Half-sandwich titanocenes used as catalysts for the polymerization of styrene

Fluorinated half-sandwich metallocenes show an increase in activity, compared to chlorinated compounds [14]. The polymerization was carried out within a temperature range of 10 to 70 °C.

Fig. 5 shows the polymerization rate of styrene by some selected half-sandwich titanocenes.

After about 10 minutes, the catalysts reach the maximum on activity. The differences between the average activities between chlorinated and fluorinated compounds can be seen in Table 3.

Under the used conditions, PhCpTiCl$_3$ is the most active catalyst. The fluorinated complexes are about three times more active than the chlorinated ones. The fluorine atom stabilizes the active site and by this increases the number of active centers.

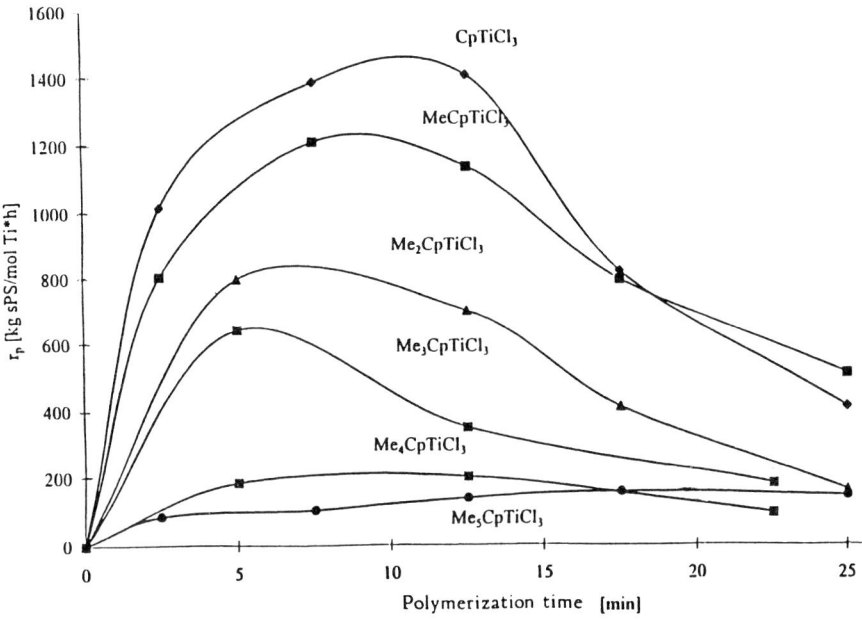

Fig. 5 Styrene polymerization rate in dependence of polymerization time for different catalysts, conditions see Tab. 3

There is another advantage of the fluorinated catalysts. They are more active by low aluminoxane/titanium ratios (Fig. 6).

The activity maximum for $CpTiF_3$ is obtained by an Al/Ti ratio of 200 while for $CpTiCl_3$ it has to be more than 1000. Using a low Al/Ti ratio, the costs especially for MAO can be decreased drastically. The activity of all investigated catalysts increases linary by the styrene concentration.

In Table 3 the molecular weights and the melting points can be found of the obtained syndiotactic polystyrenes. The molecular weights are independent on the polymerization time.

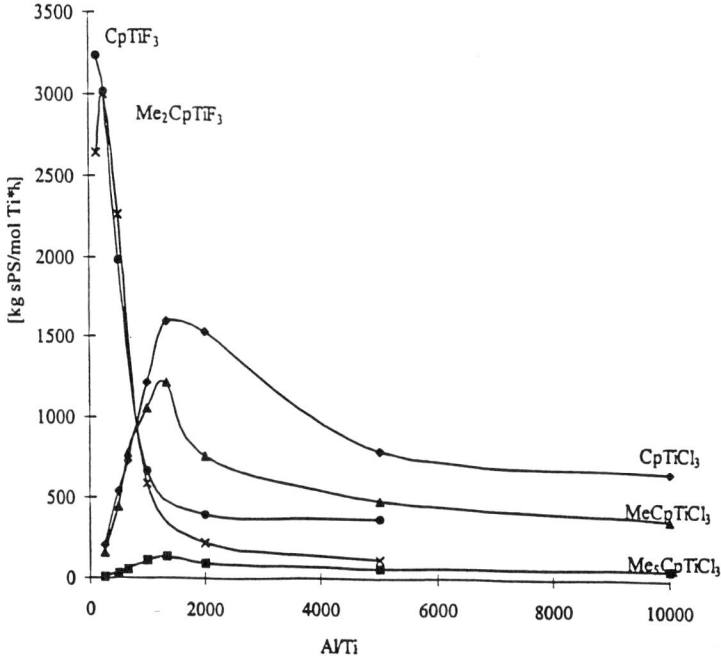

Fig. 6 Influence of the Al/Tiu molar ratio for the styrene polymerization by different half-sandwich titanocenes; reaction time = 20 min.; further conditions see Tab. 3

Table 3 Activities (kg sPS/mol Ti · h, molecular weights and melting points (°C) of different half-sandwich titanocenes for the syndiospecific polymerization of styrene by 30 °C, molecular weights and melting points for the obtained syndiotactic polystyrenes. Polymerization time 10 min.; 25 ml toluene, 25 ml styrene, Al/TiCl = 1000, Al/TiF = 500, [Ti] = 5 · 10^{-5} mol/l; 0,145 g MAO

Catalyst	Activities	Molecular weight	Melting point
CpTiCl$_3$	1 200	224 000	259
MeCpTiCl$_3$	950	190 000	266
Me$_2$CpTiCl$_3$	780	257 000	271
Me$_3$CpTiCl$_3$	550	335 000	269
Me$_4$CpTiCl$_3$	200	387 000	270
Me$_5$CpTiCl$_3$	110	469 000	270
PhCpTiCl$_3$	3 200	280 000	266
(Me$_3$Si,MeCp)TiCl$_3$	260		
CpTiF$_3$	3 100	186 000	264
MeCpTiF$_3$	3 050	258 000	270
Me$_2$CpTiF$_3$	2 700	321 000	273
Me$_3$CpTiF$_3$	830	397 000	272
Me$_4$CpTiF$_3$	460	421 000	272
Me$_5$CpTiF$_3$	350	388 000	273
Cp*TiF2(OCOCF$_3$)	270	-	273
Cp*TiF2(OCOC$_6$F$_5$	340	-	273

With an increasing substitution of the Cp-rings of the catalyst the molecular weight of the produced polystyrene increased. The molecular weight distribution Mw/Mn is around 2 and indicates that there is only a single active site. The melting points of the polystyrenes differ between 259 and 273 °C. Higher substituted catalysts increase the syndiotacticity and give polymers with higher melting points. The maximum is almost reached by the dimethyl substituted complex.

References

(1)   W. Kaminsky (ed.), *Metalorganic Catalysts for Synthesis and Polymerization*, Springer press Heidelberg 1999, p. 1-674

(2)   N. Kashiwa, J.I. Inuta, *Catalysis Surveys Japan*, **1**, 125 (1997)

(3)   W. Kaminsky, *J. Chem. Soc., Dalton Trans* 1413 (1998)

(4)   J.A. Ewen, R.L. Jones, A. Razavi, *J. Am. Chem. Soc.* **110**, 6255 (1988)

(5)   J.A. Ewen, M.J. Elder, R.L. Jones, L. Haspeslagh, J.L. Atwood, S.G. Bott, K Robinson, *Makromol. Chem. Macromol. Symp.* **48/49**, 253 (1991)

(6)   W. Spaleck, M.Antberg, V. Dolle, R. Klein, J. Rohrmann, A. Winter, *New J. Chem.* **14**, 499 (1990)

(7)   W. Kaminsky, *Macromol. Chem. Phys.* **197**, 3907 (1996), W. Kaminsky, R. Werner in Ref. 1, p. 170

(8)   W. Spaleck, W.A. Hermann, J. Rohrmann, E. Hertweck, *Angew. Chem. Int. Ed. Engl.* **29**, 1511 (1989)

(9)   N. Ishihara, M. Kuromoto, M. Uol, *Macromolecules* **21**, 3356 (1988)

(10)   A. Zambelli, L. Olivia, C. Pellecchia, *Macromolecules* **22**, 2129 (1989)

(11)   J.C.W. Chien, S. Dong, *Polym. Bull.* (Berlin) **29**, 515 (1992)

(12)   C. Pellecchia, D. Pappalardo, L. Oliva, A. Zambelli, *J. Am Chem. Soc.* **117**, 6593 (1995)

(13)   W. Kaminsky, S. Lenk, *Macromol. Chem. Phys.* **195**, 2093 (1994)

(14)   W. Kaminsky, S. Lenk, V. Scholz, H.W. Roesky, A. Herzog, *Macromolecules* **30**, 7647 (1997)

Macromol. Symp. 159, 19–26 (2000)

Characterization of Polypropylenes Prepared with Bis-Indenyl Type Metallocene and MgCl₂-Supported TiCl₄ Catalyst Systems

Norio Kashiwa*, Jun-ici Imuta, Toshiyuki Tsutsui, Shun-ichi Hama, Shin-ichi Kojoh

Organo-metal Complexes Catalization Laboratory, Mitsui Chemicals, Inc., 580-32 Nagaura, Sodegaura-City, Chiba 299-0265, Japan

SUMMARY: The homo-polypropylene: m-PP, prepared with *rac*-Me₂Si[2-*n*-Pr-4-(9-Phenanthryl)-Ind]₂ZrCl₂, showed 99.6 % of [mmmm] and 162.8 °C of melting temperature (Tm). This polymer was compared by TREF analysis with the homo-polypropylene: Ti-PP, which was produced by our latest MgCl₂-supported TiCl₄ catalyst system and showed 99.0 % of [mmmm] and 165.7 °C of Tm. It was indicated that m-PP has narrower stereoregurality distribution than Ti-PP and Tm of the fraction eluted from m-PP at the highest temperature range is 163.6 °C, while that from Ti-PP reaches to 167.3 °C. The characters and advantages of each polymer are discussed on the basis of these results. In addition, an advantage of this metallocene was made clear by characterization of polypropylenes copolymerized with ethylene.

Introduction

Since the discovery of stereospecific polymerization of propylene with the C₂-symmetrical metallocene catalyst[1], much effort has been invested for the molecular design of its ligands[2,3] and the optimization of the polymerization conditions[3,4] in order to enhance the stereospecificity. Consequently, the stereospecificities of the developed metallocene catalyst systems[2,3] have reached to a level comparative to the commercial MgCl₂-supported TiCl₄ catalyst systems. These catalyst systems have contributed to the enhancement of the regiospecificity as well as the stereospecificity, but their regiospecificities are still lower than those of the commercial MgCl₂-supported TiCl₄ catalyst systems which forbid chain-propagation reaction after regioirregular insertion of propylene monomer by the formation of dormant sites[5]. Therefore, further investigations have been continuing without a break.

          CCC 1022-1360/00/$ 17.50+.50/0

On the other hand, further improvement of the stereospecificities of the commercial MgCl$_2$-supported TiCl$_4$ catalyst systems is desired for the application of polypropylene (PP) to the automobile industry.

Under the circumstances, a comparison of the latest metallocene catalyst system with the latest MgCl$_2$-supported TiCl$_4$ catalyst system would be significant as a step toward an ideal catalyst system for the PP without any defects. The TREF analysis is an useful method for such a comparison[6]. Besides, there have been a number of reports where the TREF analysis gave a clear relation between the characteristics of the obtained PP and the natures of active sites of the MgCl$_2$-supported TiCl$_4$ catalyst system[7-10].

In this paper, the TREF analysis is carried out to compare homo-PPs produced by the latest metallocene and MgCl$_2$-supported TiCl$_4$ catalyst systems. In addition, PPs copolymerized with minor amount of ethylene (random-PPs) by these catalyst systems are characterized in order to discuss the differences in the natures of these catalyst systems.

Experimental

Catalyst

rac-Me$_2$Si[2-*n*-Pr-4-(9-Phenanthryl)Ind]$_2$ZrCl$_2$ (metallocene-A) was used as the latest metallocene catalyst because it shows the highest stereospecificity among the enormous examined metallocene catalysts in our laboratory. The structure of metallocene-A is shown in Fig.1.

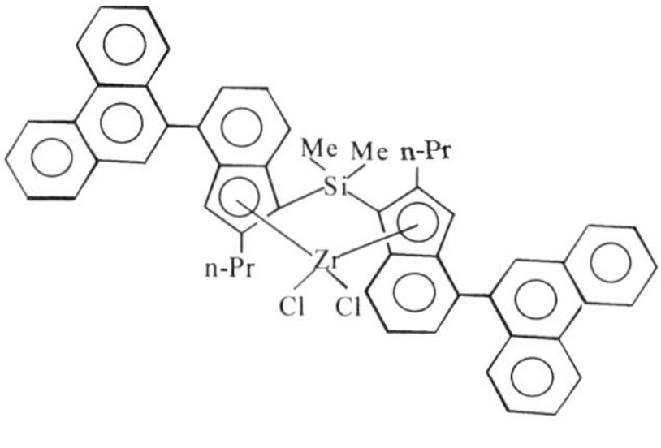

Fig. 1: Structure of metallocene-A (*rac*-Me$_2$Si[2-*n*-Pr-4-(9-Phenanthryl)Ind]$_2$ZrCl$_2$).

The latest MgCl$_2$-supported TiCl$_4$ catalyst system (Mg/Ti-A), which has been developed in our laboratory and shows higher stereospecificity than any commercially available MgCl$_2$-supported TiCl$_4$ catalyst systems, was used for the polymerizations.

Polymerization

Homo-polymerization of propylene with metallocene-A was carried out as follows. In a 2 L glass autoclave, 1 L of toluene was added and the system was charged with propylene. After cooling to 0 °C, a toluene solution dissolving 1.0 μmol of metallocene-A and 0.9 mmol of i-Bu$_3$Al was further added. Finally, 2.0 μmol of Ph$_3$CB(C$_6$F$_5$)$_4$ was added and the polymerization was carried out under atmospheric pressure for 30 min. During the polymerization, propylene and hydrogen were supplied continuously at the rates of 150 L h^{-1} and 3 L h^{-1}, respectively. At the end of the polymerization, the feed of propylene was stopped and a small amount of methanol was added. The whole product was then poured into 4 L of methanol containing a small amount of HCl. The obtained polymer was filtered, washed with excess amount of methanol and vacuum-dried at 80 °C for 12 h.

Homo-polymerization of propylene with Mg/Ti-A as follows. Into a 1 L stainless-steel autoclave, 400 mL of heptane was added and the system was charged with propylene. Then 0.4 mmol of Et$_3$Al, 0.4 mmol of an external donor and 8.0 μmol in terms of Ti (8.0 μmol-Ti) of Mg/Ti-A were added in this order at 60 °C. Next, 100 mL of hydrogen was added, and the system was pressurized to 0.49 MPa by propylene and heated to 70 °C. The polymerization was carried out at 70 °C for 1 h and propylene was fed continuously to keep 0.49 MPa. After the polymerization, the resulting slurry was filtered to separate into a powder and a liquid phase portion. The powder was washed with hexane and vacuum-dried at 80 °C for 12 h. The liquid phase portion was concentrated to obtain the heptane-soluble polymer.

The production of random-PP with metallocene-A was carried out as follows. Metallocene-A was supported on SiO$_2$ with methylaluminoxane (MAO) by pre-polymerization of propylene. The supported catalyst was fed into a sequential reactor at 0.035 mmol-Zr h^{-1} with MAO supplied at 10 mmol-Al h^{-1}. Propylene, ethylene and hydrogen were fed at the rates of 18 kg h^{-1}, 0.8 kg h^{-1} and 3.5 L h^{-1}, respectively. Polymerization was carried out at 60 °C for 2 h.

The production of random-PP with Mg/Ti-A was carried out as follows. Into a 2 L stainless-steel autoclave, 400 g of propylene, 4 L of ethylene and 4.5 L of hydrogen were added. Then 0.3 mmol of Et$_3$Al, 0.3 mmol of an external donor and 1.5 μmol-Ti of Mg/Ti-A were added in this order at 60 °C. The system was heated to 70 °C and the polymerization was carried out at that temperature for 30 min.

Polymer Characterization

The TREF analysis was carried out as follows. A stainless-steel column (21.5 mm in diameter and 150 mm in length) was packed with glass beads of 100 μm in diameters and maintained at 145 °C. Then 7.5 mL of o-dichlorobenzene (ODCB) solution dissolving 37.5 mg of the polymer sample at 145 °C was injected into the column. The cooling step was performed at 10 °C h^{-1} from 145 °C to 25 °C and the subsequent elution step was carried out by pumping ODCB at 1.0 mL min^{-1} and raising the temperature at 15 °C h^{-1} from 25 °C to 145 °C. The amount of eluted polymer was monitored for every 2.5 min by a Mercury Cadmium Telluride detector on a Nicolet FT-IR Magna-550. The whole solution was fractionated into 4 fractions and they were poured into five times volume of methanol. The precipitated polymer was collected by filtration and washed with methanol, followed by vacuum-dry.

The ^{13}C NMR, the gel permeation chromatograph (GPC) and the differential scanning calorimeter (DSC) analyses were performed according to our previous paper[5].

The portion which is soluble in decane at 23 °C (C10 Sol) was measured as follows. Into a 1 L flask, 3 g of the polymer sample was added with 20 mg of 2,6-di-t-butyl-4-methylphenol and 500 ml of decane. The mixture was heated to 150 °C in order to dissolve the polymer sample completely. The obtained solution was cooled to 23 °C during 8 h and kept at that temperature for 8 h. The resulting slurry was filtered and the liquid phase portion was vacuum-dried until it reached constant weight. The percentage of thus-obtained constant weight in the weight of the polymer sample is C10 Sol.

The ethylene content was measured according to the literature[11] and the melt flow ratio (MFR) was measured according to ASTM D1238.

Results and Discussion

The ^{13}C NMR analysis was carried out for the overall homo-PP: m-PP, prepared with metallocene-A, and the overall homo-PP: Ti-PP, prepared with Mg/Ti-A. The [mmmm] of m-PP reached to 99.6 %, while that of Ti-PP was 99.0 %. Then 0.10 mol% of 2,1-inversion was detected in m-PP in contrast with no detection in Ti-PP. It would interpret the lower Tm of m-PP than the expectation from the [mmmm] as shown in Table. 1.

The TREF analysis for m-PP and Ti-PP was carried out to discuss the differences in their characteristics. The varying amount of polymer eluted from each sample as a function of elution temperature is shown in Figure 2. While m-PP was eluted in the temperature range of 105 - 135 °C, Ti-PP was in the range of 110 - 140 °C. In both cases, the elution of polymer was negligible at temperatures below their characteristic

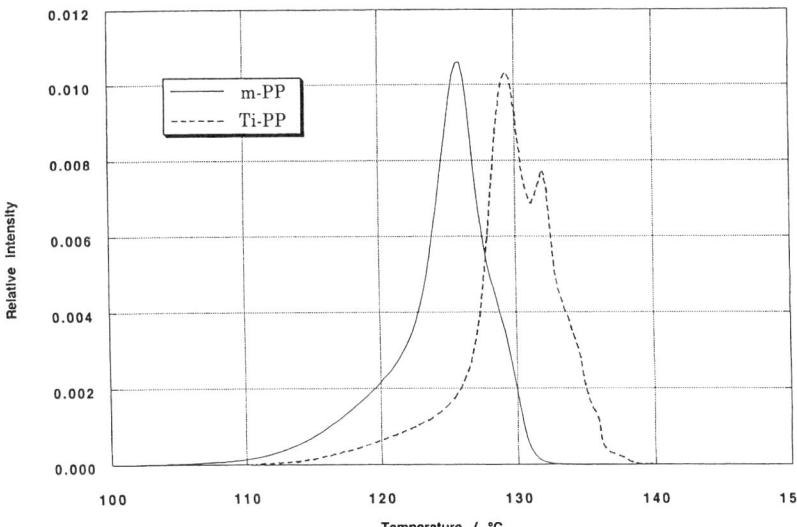

Fig. 2: TREF profiles of m-PP and Ti-PP; m-PP: homo-PP prepared with metallocene-A, Ti-PP: homo-PP prepared with Mg/Ti-A.

temperature ranges, indicating that the yield of atactic PP is too small to detect. Figure 2 shows that the elution curve of Ti-PP has two peaks at least in contrast to a single peak of m-PP. This suggests that the Mg/Ti-A and the metallocene-B would involve heterogeneous and homogeneous active sites, respectively.

In the measurement of TREF, both samples were fractionated into 4 fractions in order to analyze with DSC and GPC, and the results are shown in Table 1. Both the Tm and Mw of each fraction increased with raising the elution temperature, in which the extent of the increase for Ti-PP was much greater than that for m-PP. It is likely that the higher stereospecific active sites produce homo-PP of the higher molecular weight, especially in the use of Mg/Ti-A. Comparing fractions eluted at lower temperature (Fr.3) and higher temperature (Fr.4), the Tm increased from 166.4 °C to 167.3 °C and the Mw increased from 440,000 to 700,000 in the case of Ti-PP, while the Tm (from 163.3 °C to 163.6 °C) and the Mw (from 140,000 to 180,000) changes occurred to smaller extent in the case of m-PP. Thus, the microstructures of Fr. 3 and Fr. 4 from Ti-PP would be significantly different, while those from m-PP would be substantially the same.   It is

Table 1.   TREF, DSC and GPC analyses of m-PP (homo-PP with metallocene-A) and Ti-PP (homo-PP with Mg/Ti-A).

Sample		Elution Temp. / °C	Proportion / wt %	Tm / °C	Mw / 10^5	Mw/Mn
m-PP	whole	-	100	162.8	1.4	2.1
	Fr.1	105 - 115	3.3	-	-	-
	Fr.2	115 - 123	24.3	162.0	0.7	1.7
	Fr.3	123 - 127	49.6	163.3	1.4	1.7
	Fr.4	127 - 135	22.8	163.6	1.8	1.7
Ti-PP	whole	-	100	165.7	4.9	4.0
	Fr.1	110 - 123	7.2	160.4	-	-
	Fr.2	123 - 128	17.8	164.1	1.7	2.1
	Fr.3	128 - 131	40.3	166.4	4.4	2.5
	Fr.4	131 - 140	34.7	167.3	7.0	2.3

noteworthy that the Tm of Fr. 4 from Ti-PP is obviously higher than that of the whole sample which shows 99.0 % of [mmmm] and no regioirregularity. Although the [mmmm] of this fraction was not less than 99.7 %, the value could not be determined precisely with our current accuracy of ^{13}C NMR analysis.

The results of characterization of random-PPs prepared with metallocene-A and Mg/Ti-A were summarized in Table 2. Commonly, the portion which is soluble in decane at 23 °C (C10 Sol) is considered as the polymer containing much more ethylene comonomer than the overall polymer. Therefore, the higher value of C10 Sol of the random-PP with Mg/Ti-A at the almost same Tm as metallocene-A would show the broader distribution of ethylene content in the resulting polymer than metallocene-A. Besides, the higher ethylene content of the random-PP with Mg/Ti-A at the almost same Tm as metallocene-A would indicate that metallocene-A is more effective than Mg/Ti-A for lowering Tm of the resulting random-PP by the copolymerized ethylene unit.

Table 2. Results of characterization of random-PPs prepared with metallocene-A and Mg/Ti-A.

Catalyst	Tm / °C	C10 Sol / wt%	ethylene content / mol%	MFR / dg min^{-1}
metallocene-A	132	1.4	3.2	8.8
Mg/Ti-A	131	6.7	5.5	3.6

Conclusions

The TREF analysis indicated that the homo-PP prepared with metallocene-A had narrower distribution of stereoregularity than Mg/Ti-A. While the Tm of the homo-PP eluted at the highest temperature range in the case of metallocene-A was 163.6 °C, that of corresponding fraction with Mg/Ti-A reached up to 167.3 °C. Those would show each advantage of metallocene-A and Mg/Ti-A, namely homogeneity and regiospecificity, respectively. The characterization of the random-PPs with both the catalyst systems made an advantage of metallocene-A clear. This catalyst could lower the Tm of the random-PP by the smaller amounts of the ethylene content and C10 Sol than Mg/Ti-A. In view of these evidences, there may be two approaches to the development of an ideal catalyst. One is the enhancement of the regiospecificity of metallocene-A and another is to unify the plural active sites of Mg/Ti-A into the highly stereospecific active site that can produce such a PP eluted as Fr. 4 from Ti-PP.

References

1. W. Kaminsky, K. Kulper, H. H. Brintzinger, F. R. W. P. Wild, *Angew. Chem., Int. Ed. Engl.* **24**, 507 (1985)
2. W. Spalek, F. Kuber, A. Winter, J. Rohrmann, B. Bachmann, M. Antberg, V. Dolle, E. F. Paulus, *Organometallics* **13**, 954 (1994)
3. H. Deng, H. Winkelbach, K. Taeji, W. Kaminsky, K. Soga, *Macromolecules* **29**, 6371 (1996)
4. A. Toyota, T. Tsutsui, N. Kashiwa, *J. Mol. Catal.* **56**, 237 (1989)
5. S. Kojoh, M. Kioka, N. Kashiwa, M. Itoh, A. Mizuno, *Polymer* **36**, 5015 (1995)
6. A. Mizuno, T. Abiru, M. Motowoka, M. Kioka, M. Onda, *J. Appl. Polym. Sci: Appl. Polym. Symp.* **52**, 159 (1993)
7. M. Kakugo, T. Miyatake, Y. Naito, K. Mizunuma, *Macromolecules* **21**, 314 (1988)
8. M. Kioka, H. Makio, A. Mizuno, N. Kashiwa, *Polymer* **35**, 580 (1994)
9. J. Xu, F. Feng, S. Yang, Y. Yang, X. Kong, *Polym. J.* **29**, 713 (1997)
10. I. Mingozzi, G. Cecchin, G. Morini, *Int. J. Polym. Anal. & Charact.* **3**, 293 (1997)
11. M. Kakugo, T. Naito, K. Mizunuma, T. Miyatake, *Macromolecules* **15**, 1150 (1982)

Macromol. Symp. **159**, 27–34 (2000)   27

Preparation of New Biomedical Materials and their Application in Clinics

Nobuo Nakabayashi

Institute of Biomaterials and Bioengineering, Tokyo Medical and Dental University

Kanda-Surugadai, Tokyo 101-0062, Japan

SUMMARY: Biomaterials - tissues interaction is important to study in biomaterials science. The information is indispensable to make medical devises and artificial organs and to predict their performance. It is also very useful to consider a hypothesis to design new biomaterials. New materials have brought big progress in the society as we know. There are few biomaterials specially designed to use in biomedical fields. The most important effort must be preparation of biocompatible materials, that must be essential to develop new type high performance devices and artificial organs. Preparation of new dental biomaterials used in bonding of prostheses to dentinal tissues that require fundamental change in modern dentistry and a new methacrylate, MPC, to develop promising several kinds of biomaterials with unusually excellent biocompatibility and functions are going to present. Topics in tissue engineering are also discussed.

Introduction

Biomedical materials are very important today in polymer science and technology. Many polymers have widely been used in medical fields directly and/or indirectly. Unfortunately, they are mainly conventional products, not specially prepared for biomedical application. The most important factor in biomaterials is biocompatibility. It has several meanings. Before discussing biocompatibility, non-toxicity to our body must be guaranteed. Some believe non-toxicity is same as biocompatibility. The preparation of biocompatible biomaterials is a key step. In artificial kidney devices, blood shunts are required to transfer blood into a hemodialyzer for purification. Infection or clotting at the interface between skin or blood vessels and shunts has fostered the use of fistulas into which needle insertion is made. Artificial heart must be connected with arteries and veins. Interfaces between artificial materials and tissue or blood have always been troublesome. They are clotting, infection and rejection, etc. Unfortunately we have not had reliable ideas for solving these problems.

Application of biomaterials

There must be described many biomaterials used in disposable and implantable medical devices. Moreover, each of them possesses their own unique problems to be resolved.

          CCC 1022-1360/00/$ 17.50+.50/0

Fig. 1 represents where biomaterials are used as artificial organs to support defects of our body. Several medical devises are widely used for medical treatments as shown in Table 1. Most of their raw materials are conventional commercialized polymers. They are plasticized polyvinyl chloride, polypropylene, polyester, cellulose, polysulfone, polyurethanes, etc.

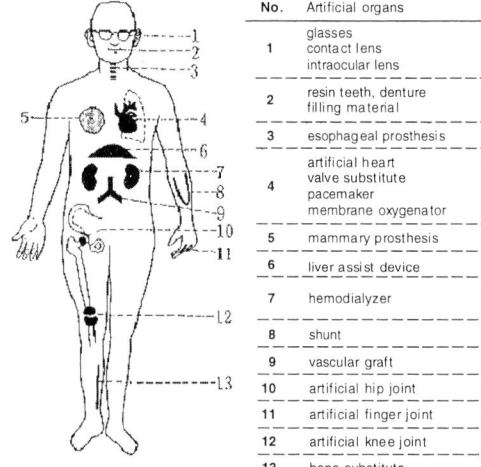

No.	Artificial organs	Biomaterial
1	glasses contact lens intraocular lens	CR-39,MR-6,PMMA PMMA, PHEMA, MPC polymer PMMA
2	resin teeth, denture filling material	PMMA methacrylates & fillers
3	esophageal prosthesis	polyethylene/natural rubber
4	artificial heart valve substitute pacemaker membrane oxygenator	segmented polyurethanes carbon coated titanium, gluteraldehyde-preseved polyurethane porous polypropylene, silicone (extracorporal)
5	mammary prosthesis	silicone
6	liver assist device	coated charcoal, porous polymer-beads
7	hemodialyzer	cellulose, cellulose acetate, EVAL, PMMA, polysulfone
8	shunt	silicone/PTFE
9	vascular graft	polyester (knit), PTFE
10	artificial hip joint	metal alloys/UHMWPE
11	artificial finger joint	silicone
12	artificial knee joint	metal alloys/UHMWPE
13	bone substitute	hydroxyapatite, alumina, titanium alloy

Fig. 1: Biomaterials used in artificial organs

Table 1. Products of biomedical devises in Japan (1996)

Device	Product (pieces x 10^4)	Biomaterial
disposable needle	279 198	stainless steel
disposable syringe	207 428	polypropylene (PP)
disposable catheter and circuit		
digestive, respiratory, ureteral	8 296	plasticized PVC
vessel	7 765	plasticized PVC, silicone polyurethane
catheter for hemodialysis	3 787	PTFE
blood taking set	66 270	silicone, plasticized PVC
blood bag	753	plasticized PVC
blood infusion set	1 059	plasticized PVC
transfusion solution set	60 602	plasticized PVC
extracoporeal circuit	4 242	plasticized PVC
hollow fiber hemodialyzer	4 435	cellulose, cellulose acetate, polysulfone, PMMA, EVAL polycarbonate (housing)
artificial joint	12.2 (import 90 %)	metal
	3.3 (import 5 %)	ceramics
intraocular lens	75	PMMA
membrane oxygenator	15	porous PP, silicone

They must be sterilized either by heat, steam, chemicals or irradiation with UV light or γ-ray. Before making medical devices commercially available, the approval must be taken from the regulatory agency, the FDA , the Ministry of Public Welfare, etc., and they have to be produced following approved processes exactly.

There are two major classes of tissue, hard and soft, in the body. When ideal biocompatible materials is implanted, the molecules of the implant and those of the host tissues may mix homogeneously at the interface as in a blend polymer. This kind of biocompatible polymer has been sought for a long time. The hard tissues could divide into two, one is tooth that does not regenerate and the other is bone. Connection of biomaterials with tooth could be possible by developing good adhesives, as the tissue could not fill the space. Fortunately, in the case of bone, the space could be filled by newly formed bone when the materials do not give adverse effect on the regeneration. Connection of polymers to soft tissues is very difficult and we have to challenge to provide good resolution. Such biomaterials that could encourage tissue- or cell-growth and accept their invasion into them could be a candidate. Examination of soft tissue compatible materials is particularly difficult because of differences in mechanical properties between implants and the tissues. It is very hard to distinguish the lack of compatibility from mechanical failure at the interface when the implant ends in failure. Stress concentration problems at implant interface would be severe in cases of soft tissue biocompatibility. We have been working hard to prepare good blood compatible materials to support artificial heart programs. Unfortunately, they became available without development of new blood compatible materials. These explanations could be illustrated as Fig. 2.

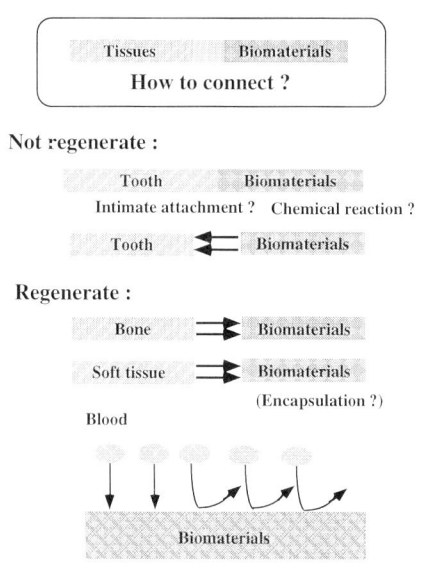

Fig. 2 : How to connect tissues and biomaterials

Dental polymers[1)]

Dental hard tissues do not regenerate. When they get defect, we have to use biomaterials to restore the function and esthetics. So it has been believed that dental biomaterials could support development of dental treatments for long time. And dentists try to replace natural tissues with artificial materials. But enamel is much more superior to any artificial materials. They should have considered properties and function of enamel itself more carefully. Once burs remove enamel for a dental treatment, dentin and pulp are exposed to several stimuli. And it is very difficult for them to survive longer that could be modern dental treatments. It resulted in promoting loss of tissues but they believed they treated tooth to recover the defects.

It was hypothesized that when adhesive technology could be introduced, they could improve dental treatments and better clinical results could be expected. This was not bad idea but introduction of conventional adhesive technology in dentistry was not successful. We had to think about the bonding substrate being a natural tissue, not artificial surfaces. The important point was how to get adhesion to dental hard tissues, kind of living tissues. Connection of biomaterials with natural tissues is essential in biomaterials field but there is not reliable bonding mechanism available even today except bonding to tooth substrates. Nakabayashi found in 1982 that hybridization of dental hard tissues could unite dentin with polymers (see Fig. 2).

Bonding to dentin is taken place when the tissue surface has improved to have good affinity with artificial materials such as acrylic polymers. Then we could bond acrylic polymers to the modified dentin surface. Generally, soft tissues do not accept low molecular weight man-made compounds and must be irritated by their impregnation. But, dental hard tissues are not so active and could permit monomer impregnation. Then, we could polymerize the monomer in situ and the hybridized dentin, a molecular mixture of dentinal substances including collagen fibrils and polymer chains, could be prepared. New methacrylates, which could promote monomer diffusion into dentin, were synthesized as shown in Fig. 3. They have both hydrophilic and hydrophobic group in the molecule. It means they have good affinity with dentinal hard tissues as the

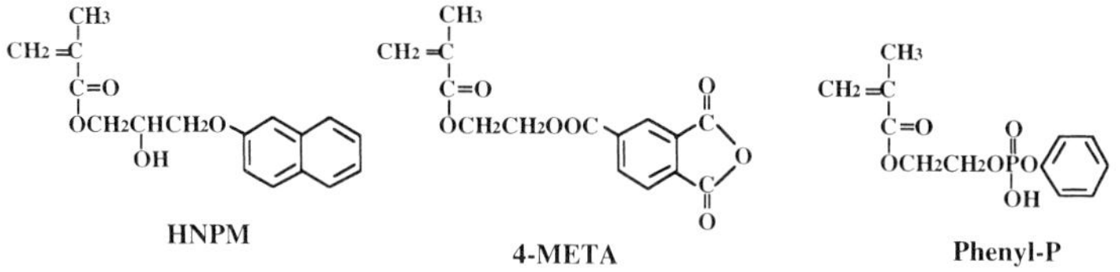

Fig. 3 : Hydrophobic and hydrophilic methacrylates for bonding agents to tooth substrates

tissue accepts their impregnation. We could say they are biocompatible methacrylates.

Dentin is mainly composed of hydroxyapatite that is demineralized by acids and collagen fibrils that are degraded by sodium hypochlorite. Acrylic polymers are stable against both acids and sodium hypochlorite in the same condition. Then, we could differentiate three of them even when they are mixed at molecular level in the hybridized dentin by comparison of morphological changes of polished, demineralized and then proteolytic-degraded surfaces with SEM examination after soaking samples in 6 mol/L HCl for 30 sec and then in 1 % NaOCl for 10, 30 and/or 60 min. The good hybridized dentin could resist well against both acidic and proteolytic attack. The chemical reactions must be akin to those of caries formation. This means the hybridized dentin could protect dentinal caries that has been impossible before in dentistry.

Usually dentin requires epoxy resin embedding before ultra-thin cross-sectioning. It reinforces the substrate. Epoxy embedding is a kind of hybridization of dentin with the epoxy resin. When we could prepare ultra-thin cross-section directly from the hybridized dentin with acrylic polymers (adhesive resins), we could conclude the hybridization was carried out well. It also suggests that the bonded dentin specimens could well

Fig. 4 : TEM picture of a resin-dentin bond created *in vivo*. The hybrid layer (H) appears to be well infiltrated with 4-META/MMA-TBB resin (R). d: Mineralized dentin.

resist shear stress loaded by cutting instrument. The hydroxyapatite crystallites in the hybridized dentin (H in Fig. 4) could resist against HCl demineralization. This suggests that hybridized dentin is an impermeable membrane created in the subsurface of dentin that could protect pulp from invasion of several stimuli. We have not had such excellent biomaterials before.

Development of MPC surfaces for biomedical application[2)]

Improvement of blood compatibility has been top interest for long time but we do not have reliable materials even today. Hydrophilic surfaces covered with hydrophilic moieties like polyethylene glycol, heparinized surfaces and poly (styrene-block-HEMA) were promising surface. It was hypothesized that polymers having good affinity with phospholipids could adsorb them on the surface and might show new interesting properties as biomaterials. Phosphorylcholine is a major phospholipid in many cells and tissues. And 2-methacryloyloxyethyl phosphorylcholine (MPC, Fig. 5) was prepared. It was found that MPC copolymers have several functions good for biomaterials, such as long term blood compatibility (Fig. 6)[3)], inhibition of proteins adsorption and increasing their stability, good permeability for mass transport, transparent hydrogel and good to keep surface moist, etc. MPC can copolymerize with many conventional monomers and it is easy to prepare desired biomaterials with multi-characteristics. Modification of fabricated devises is also possible (Fig. 6). The most interesting points were that copolymers accumulate the phospholipid on the surface and the accumulated could construct biomembrane-like surface, which possesses same melting point as the liposome. It is a kind of biomimetic membrane surface (Fig. 7).

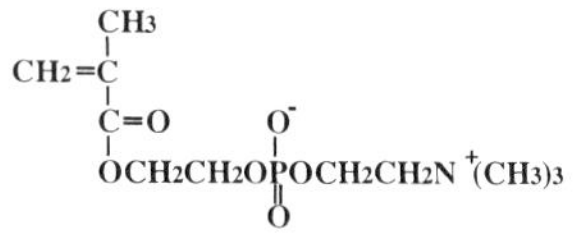

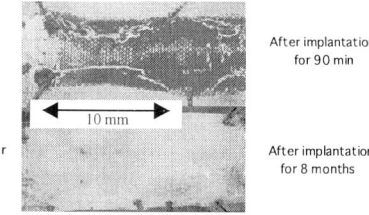

$$CH_2=\overset{\overset{\displaystyle CH_3}{|}}{\underset{\underset{\displaystyle OCH_2CH_2O\overset{O}{\underset{O}{P}}OCH_2CH_2N^+(CH_3)_3}{|}}{\underset{|}{C}}}$$

Fig. 5 : MPC

SPU — After implantation for 90 min

10 mm

SPU/MPC polymer blend — After implantation for 8 months

Fig.6 : Optical microscopic picture of small diameter blood vessels.

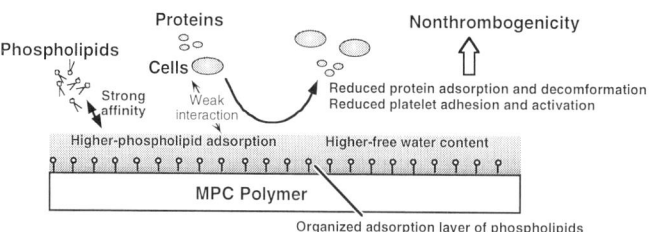

Fig.7 : Illustration of biomimetic membrane on MPC polymer surface

MPC is useful in contact lenses and their care, hemodialysis and oxygenator membranes effective even absence of anticoagulants, long lasting glucose monitoring biosensors (Fig. 8) , surface coatings for several catheters and devises, prevention of protein accumulation, lubrication, carriers for drug delivery systems and increasing stability of enzymes, etc.

Biomaterials and non-invasive surgery

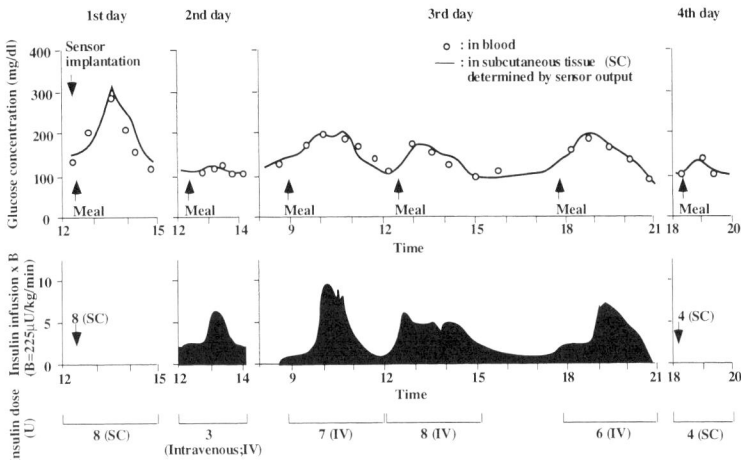

Fig. 8 : Continuous monitoring of subcutaneous tissue glucose concentrations and blood glucose regulation in an insulin requiring NIDDM patient with a ferrocene-mediated needle-type glucose sensor covered with MPC polymer membrane.

Recently, less-invasive surgery is required in many cases and biomaterials could contribute on the developments. It could improve patient quality of life very much. Using several new functional catheters such as balloon catheters could develop new surgical technology.

Biomaterials and tissue engineering

Tissue engineering technology is very important for biomaterials field to develop bioartificial organs. Preparation of tissue reconstruction would be possible by combining gene-technology. An artificial skin could be prepared by this technology using the patient skin elements. Small diameter vascular grafts are fabricated also. It has been difficult to prepare the graft out of biomaterials as thrombosis occludes easily the small open space where blood flows before neointima could cover the surface. Hybridization of small vascular graft with endothelial cells, which secrete heparin, is promising way (Fig. 9)[4].

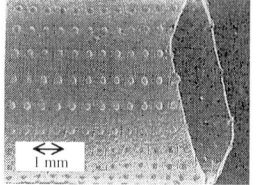

Another important targets must be artificial organs that require solutes transport mechanism. Typical examples are hepatic and pancreatic assist devices. In these cases,

Fig. 9 : Gelatin coated polyurethane substrate with pores to prepare small diameter vascular graft by tissue engineering.

a combination of suitable biomaterials with tissues or cells such as hepatic or islet cells could afford good resolution. We could call them bioartificial organs (devices) or hybrid artificial organs. Oxygen, carbon dioxide, nutrients and cell products must be transported through the biomaterials. We have to make research how to prepare three-dimensional matrix out of biomaterials that make tissues or cells cultivation possible. Vascularization between cells and biomaterials are encouraged for their transportation. Their immunologic interactions with patients are also considered. In this sense, we have to develop new multifunctional and high performance biomaterials before tissue engineering possible.

References

1. N. Nakabayashi, D. H. Pashley, Hybridization of Dental Hard Tissues, Quintessence Publishing, Tokyo, Berlin, Chicago 1998
2. K. Ishihara, *Trends Polym. Sci.* **5**, 401 (1997)
3. T. Yoneyama, K. Ishihara, N. Nakabayashi, M. Ito, Y. Mishima, *J. Biomed. Mater. Res. (Appl. Biomater.)* **43**, 15 (1998)
4. K. Doi, S. Satoh, T. Ota, T. Matsuda, *Am. Soc. Artificial Inter. Organs J.* **42**, 394 (1996)

Effect of End Group Chemistry on Surface Molecular Motion of Monodisperse Polystyrene Films

Tisato Kajiyama*, Noriaki Satomi, Yasuyuki Yokoe, Daisuke Kawaguchi,
Keiji Tanaka, Atsushi Takahara

Graduate School of Engineering, Kyushu University, 6-10-1 Hakozaki,
Higashi-ku, Fukuoka 812-8581, Japan

SUMMARY : The surface molecular motion of monodisperse polystyrene (PS) with various chain end groups was investigated on the basis of temperature-dependent scanning viscoelasticity microscope (TDSVM). The surface glass transition temperatures, T_g^ss for the proton-terminated PS (PS-H) films with number-average molecular weight, M_n of 4.9k - 1,450k measured by TDSVM measurement were smaller than those for the bulk one, with corresponding M_ns, and the T_g^ss for M_n smaller than ca. 50k were lower than room temperature (293 K).   In the case of M_n = ca. 50k, the T_g^ss for the α,ω-diamino-terminated PS (α,ω-PS(NH$_2$)$_2$) and α,ω-dicarboxy-terminated PS (α,ω-PS(COOH)$_2$) films were higher than that of the PS-H film.   On the other hand, the T_g^s for the α,ω-perfluoroalkylsilyl-terminated PS (α,ω-PS(SiC$_2$C$^F_6$)$_2$) film with the same M_n was much lower than those for the PS films with all other chain ends.   The change of T_g^s for the PS film with various chain end groups can be explained in terms of the depth distribution of chain end groups at the surface region.

Introduction

Glass transition temperature, T_g is the fundamental parameter of polymeric solids which is closely related to the state of thermal molecular motion.   In the case of ultrathin polymer films, the glass transition behavior has been evaluated based on spectroscopic ellipsometry[1], X-ray reflectivity[2], and Brillouin light scattering[3]. These measurements revealed that the magnitude of T_g for the ultrathin film depended on the film thickness and the state of interaction between polymer chain and substrate.   Since these measurements were based on the temperature change in average physicochemical properties of polymer ultrathin films, it is difficult to measure the molecular motion of polymer film at the outermost surface region.

Scanning force microscopy (SFM) can directly evaluate the physicochemical properties at the outermost surface region on the basis of the deflection and torsion of the cantilever which reflect atomic force, adhesion, friction, magnetic or chemical

                CCC 1022-1360/00/$ 17.50+.50/0

interaction between tip and polymer surface.   In our previous study, the surface relaxation process for the monodisperse polystyrene (PS) films with different number-average molecular weights, M_ns had been investigated by lateral force microscopy (LFM)[4] and scanning viscoelasticity microscopy (SVM)[5].   From the temperature change in the shift factors used in the formation of the master curve, the apparent activation energy for the surface α_a-relaxation process of the monodisperse PS film was found to be approximately half that for the bulk one[6].   The purpose of this study is to evaluate the influence of the chain end chemistry on the magnitude of surface T_g, T_g^s for the monodisperse PS films with various M_ns on the basis of temperature-dependent SVM (TDSVM)[7].

Experimental

The polymers used in this study were monodisperse four kinds of PSs with various chain end chemistry, and were synthesized by a living anionic polymerization. Table I tabulates the M_n, the molecular weight dispersity, M_w/M_n, and the bulk T_g, T_g^b for four kinds of PSs.   The T_g^b was evaluated on the basis of differential scanning calorimetric (DSC) measurement at a heating rate of 10 K min^{-1} under a dry nitrogen purge.   The monodisperse sample was used in order to avoid the influence on

Table I. Characterizations of four kinds of PSs used in this study.

Sample	M_n	M_w/M_n	T_g^b / K
PS-H	4.9k	1.09	362
	30k	1.08	374
	54k	1.03	377
	90k	1.05	378
	250k	1.03	379
	1,450k[a]	1.06	380
α,ω - PS(NH$_2$)$_2$	23k	1.08	375
	52k	1.08	378
α,ω - PS(COOH)$_2$	12k[b]	1.09	385
	52k[b]	1.08	382
α,ω - PS(SiC$_2$C$^F_6$)$_2$	25k	1.12	373
	48k	1.17	377
	93k	1.17	379

[a]Purchased from Pressure Chemical Co., Ltd.
[b]Purchased from Polymer Source. Inc.

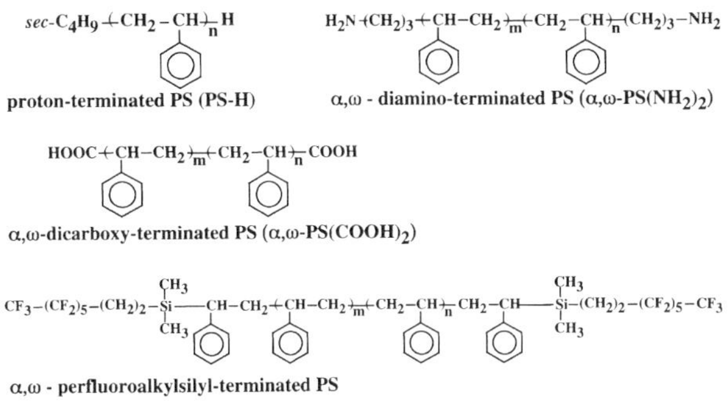

Fig. 1: Chemical structure of PSs with various chain end groups.

surface migration of lower M_n in polydisperse sample.   Fig. 1 shows the chemical structure of the PSs with various chain end groups.   The PS films were coated from a toluene solution onto cleaned silicon wafer by a spin-coating method.   The film thickness evaluated by ellipsometric measurement was ca. 200 nm.   These films were annealed at 423 K for 24 hrs under vacuum to achieve the thermal equilibrium state.   In order to investigate the surface molecular motion of these PS films, the surface dynamic viscoelastic function, in other words, the surface phase lag between dynamic deformation and force signals was measured as a function of temperature based on TDSVM.   TDSVM equipment was SPA 300HV (Seiko Instruments Industry Co., Ltd.) with an SPI 3800 controller.   TDSVM measurement was carried out at various temperatures in vacuo under a reference force of ~1 nN (repulsive force).   The modulation frequency and the modulation amplitude at the supporting part of the cantilever were 4 kHz and 1.0 nm, respectively.   A commercially available silicon nitride (Si_3N_4) tip on a rectangular cantilever with the bending spring constant of 0.09 N m^{-1} (Olympus Co., Ltd.) was used.

Results and Discussion

Fig. 2 shows the temperature dependence of the surface phase lag, δ^s between imposed stimulation displacement and detected response force signals, and the bulk tanδ for the monodisperse proton-terminated PS (PS-H) films with M_ns of 4.9k, 30k, 54k, 90k, 250k, and 1,450k.   The temperature dependence of bulk tanδ was obtained by Rheovibron (DDV-01FP, Orientec A&D Co., Ltd.) at 3.5 Hz.   Though it might be possible that the measured δ_s contains the contribution of the phase lag from the mechanical vibration system[8], the calibration has not been done at the

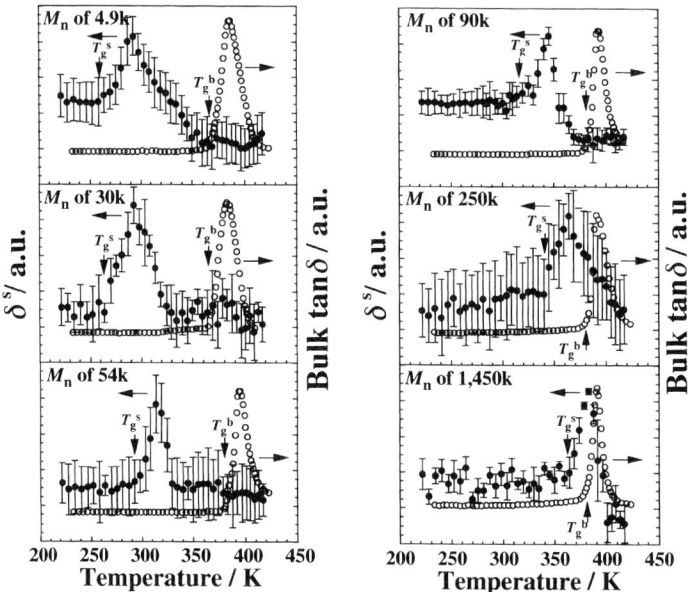

Fig. 2: Temperature dependence of surface phase lag, δ^s and bulk tanδ for the PS-H films with different M_ns.

present time because of the difficulty in determination of the calibration constant. However, it seems reasonable to compare quantitatively the temperature dependence of the δ^s and the bulk tanδ to decide the T_g^s and the T_g^b since the calibration constant might not change so much in the temperature range around these T_g^ss. Fig. 2 shows the α_a loss peaks corresponding to the surface and bulk micro-Brownian motion of polymeric chains, respectively. The peaks of δ^ss appeared in a lower temperature region than those bulk tanδs, though the δ^s and the bulk tanδ were measured at 4 kHz and 3.5 Hz, respectively. The threshold temperature, that is, the temperature at which the magnitude of δ^s or bulk tanδ starts to increase, can be empirically used as the surface or bulk T_g^s[9]. Then, in this study, this threshold temperatures of the δ^s and the bulk tanδ were defined as T_g^s and T_g^b, respectively. The T_g^s and T_g^b are shown by arrows in Fig. 2. The T_g^ss were more apparently shifted to a higher temperature with an increase in M_n than T_g^bs, as shown in Fig. 2. Therefore, it seems reasonable to conclude that the surface thermal molecular motion is more activated in comparison with the bulk one. Also, the T_g^s for the PS-H film was strongly dependent on the M_n as compared with the bulk one. These results indicate that the surface molecular motion of the monodisperse PS-H film is closely related to the concentration of chain end groups at the film surface. Since the magnitudes of the T_g^s for the PS-H films with M_n lower than ca. 50k were below 293 K, it seems reasonable to conclude that the PS-H films surface with M_n lower than ca. 50k is in a glass-rubber transition state or a rubbery state, even at room temperature, 293 K.   The T_g^s asymptotically approaches the bulk one with an increase in M_n and the T_g^s for high M_n of ca. 1,000k was slightly lower than the bulk one.   Therefore, these results apparently indicate that a reduction of T_g^s of PS-H in comparison with T_g^b may be caused by an increased free volume at the surface due to the preferential surface enrichment of chain end groups as well as the decrease in number of chain entanglement density[10].   Also, these results support our previous result that the magnitude of activation energy for the surface α_a-relaxation process at the air-polymer interface was almost half as much as that for the bulk one, maybe due to an extra freedom for the active thermal segmental motion of PS chains at the air-polymer interface[6].

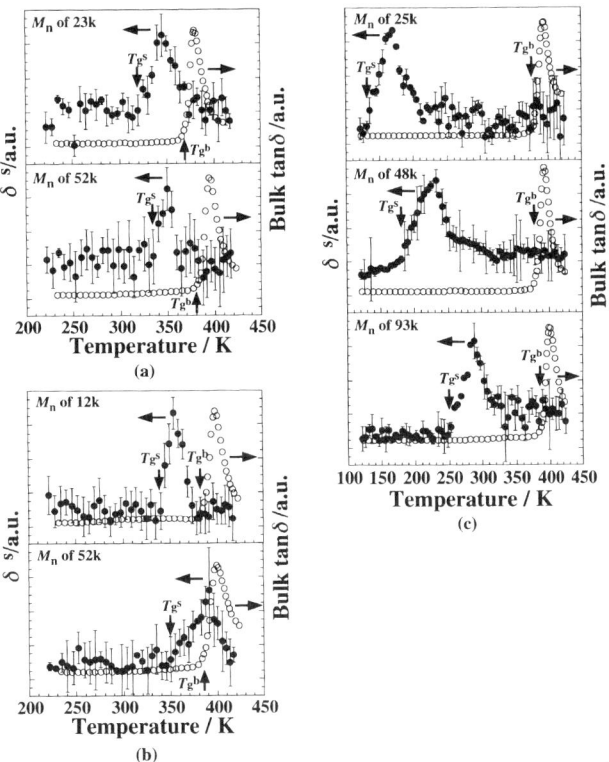

Fig. 3: Temperature dependence of δ^s and bulk $\tan\delta$ for (a) α,ω-PS(NH$_2$)$_2$, (b) α,ω-PS(COOH)$_2$ and (c) α,ω-PS(SiC$_2$C$^F_6$)$_2$ films.

The previous results revealed that the surface molecular motion of the monodisperse PS-H film was closely related to the distribution of chain end groups at the outermost surface [4-6]. However, the surface molecular motion of the PS film with different kind of chain end groups has not been evaluated yet [11]. PSs with both chain end groups of amino, carboxyl, and fluoroalkylsilyl groups were prepared, and the T_g^s's of these PSs were measured based on TDSVM. Fig. 3 shows the temperature dependence of the δ^s and the bulk $\tan\delta$ for (a) the α,ω-diamino-terminated PS (α,ω-PS(NH$_2$)$_2$) films with M_ns of 23k and 52k, (b) the α,ω-dicarboxy-terminated PS (α,ω-PS(COOH)$_2$) films with M_ns of 12k and 52k, and (c) the α,ω-perfluoroalkylsilyl-terminated PS (α,ω-PS(SiC$_2$C$^F_6$)$_2$) films with M_ns of 25k, 48k and 93k. The peaks of δ^s, which correspond to the surface α_a-relaxation process, were similarly observed on the δ^s-temperature curves of all kinds of PSs and appeared in a lower temperature region than those of bulk $\tan\delta$s, as shown in Fig. 3 (a), (b) and (c). As mentioned above, it is empirically accepted that the threshold temperature of δ^s and bulk $\tan\delta$ were decided as T_g^s and T_g^b, respectively. In the same manner as PS-H, the magnitude of the T_g^s's for the α,ω-PS(NH$_2$)$_2$, α,ω-PS(COOH)$_2$ and α,ω-PS(SiC$_2$C$^F_6$)$_2$ films was lower than the bulk one. However, these T_g^s's at the same M_n (M_n = ca. 50k) were strongly depended on the end group chemistry. At the same M_n, the T_g^s's for the α,ω-PS(NH$_2$)$_2$ and α,ω-PS(COOH)$_2$ films were higher than that for the PS-H film. On the other hand, a larger depression of T_g^s for the α,ω-PS(SiC$_2$C$^F_6$)$_2$ film than those of PSs with other chain end groups was observed.

Fig. 4 shows the molecular weight dependence of the surface and bulk T_gs for the monodisperse PS films with various chain end groups. The T_g^s's of α,ω-PS(NH$_2$)$_2$ and α,ω-PS(COOH)$_2$ were higher than those of PS-H with corresponding M_n. Since both chain end groups for α,ω-PS(NH$_2$)$_2$ and α,ω-PS(COOH)$_2$ have higher surface free energy in comparison with the main chain part, the chain end groups are depleted from the surface region. Also, because α,ω-PS(COOH)$_2$s are intermolecularly associated by hydrogen bonding, the apparent molecular weight of α,ω-PS(COOH)$_2$ increases or the networks of PS chains were formed, resulting in a decrease in surface molecular motion. Thus, it seems reasonable to conclude that the T_g^s's of α,ω-PS(COOH)$_2$ were higher than those of α,ω-PS(NH$_2$)$_2$ and PS-H at the same M_n. On the other hand, the T_g^s's of α,ω-PS(SiC$_2$C$^F_6$)$_2$ film with all M_ns were much lower than the bulk one and that for the PS-H film. Since the fluoroalkylsilyl end group is more hydrophobic than the main chain part and the *sec*-butyl and –CH$_2$-CH$_2$(C$_6$H$_5$) groups, which are the chain end group of PS-H, the surface free energy of fluoroalkylsilyl end group is lower than that of PS-H.

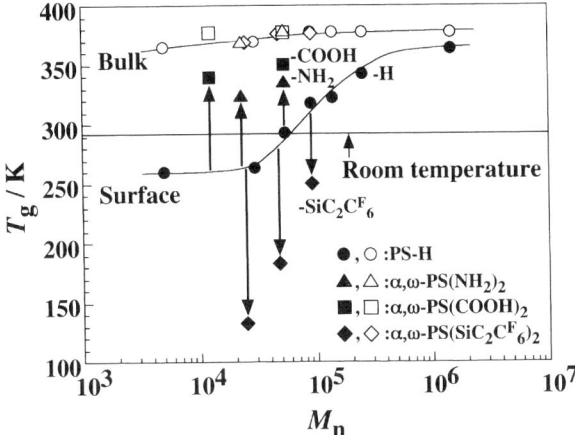

Fig. 4: Molecular weight dependence of surface and bulk T_gs for the PS films with various chain end groups. The filled and open symbols denote the data for the T_g^s and T_g^b, respectively.

Therefore, a larger depression of T_g^s for the α,ω-PS(SiC$_2$C$^F_6$)$_2$ film than that of PS-H was explained by the intensive surface localization of fluoroalkylsilyl end group compared with other kind of PSs. In the case of the α,ω-PS(SiC$_2$C$^F_6$)$_2$ film, this intensive surface localization of fluoroalkylsilyl end group in the surface layer less than $\sim$ 1 nm was confirmed on the basis of angular-dependent X-ray photoelectron spectroscopic (ADXPS) measurement.

Conclusion

The molecular weight dependences of the T_g^s for the monodisperse PS film with various chain end groups were evaluated by TDSVM measurement. The T_g^ss for the various chain end group terminated PS film were much lower than the bulk ones, and the T_g^s was strongly dependent on a difference of surface free energy between chain end group and main chain. The difference of T_g^s was explained by the surface concentration of chain end groups which is closely related to the free volume fraction.

References

1. J. L. Keddie, R. A. L. Jones, R. A. Coury, *Europhys. Lett.* **27**, 59 (1994)
2. W. J. Orts, J. H. van Zanten, W. L. Wu, S. K. Satija, *Phys. Rev. Lett.* **71**, 867 (1993)

3. J. A. Forrest, K. Dalnoski-Veress, J. R. Stevens, J. R. Dutcher, *Phys. Rev. Lett.* **77**, 2002 (1996)

4. (a) T. Kajiyama, K. Tanaka, A. Takahara, *Macromolecules* **30**, 280 (1997), (b) K. Tanaka, A. Takahara, T. Kajiyama, *Macromolecules* **30**, 6626 (1997)

5. (a) T. Kajiyama, K. Tanaka, I. Ohki, S. R. Ge, S. J. Yoon, A. Takahara, *Macromolecules* **27**, 7932 (1994), (b) K. Tanaka, A. Taura, S. R. Ge, A. Takahara, T. Kajiyama, *Macromolecules* **29**, 3040 (1996)

6. T. Kajiyama, K. Tanaka, N. Satomi, A. Takahara, *Macromolecules* **31**, 5150 (1998)

7. N. Satomi, A. Takahara, T. Kajiyama, *Macromolecules* **32**, 4474 (1999)

8. N. A. Burnham, G. Gremaud, A. J. Kulik, P. J. Gallo, F. Oulevey, *J. Vac. Sci. Technol. B* **14**, 1308 (1996)

9. N. Saitô, K. Okano, S. Iwayanagi, T. Hideshima, in: *Solid State Physics* **Vol.14**, F. Seitz, D. Turubull, Eds., Academic Press, New York and London 1963, p.343

10. H. R. Brown, T. P. Russell, *Macromolecules* **29**, 798 (1996)

11. K. Tanaka, X. Jiang, K. Nakamura, A. Takahara, T. Kajiyama, T. Ishizone, A. Hirao, S. Nakahama, *Macromolecules* **31**, 5148 (1998)

Macromol. Symp. **159**, 43–55 (2000)

Sol-Gel Derived Nanoparticles as Inorganic Phases in Polymer-type Matrices

Helmut Schmidt

Institut für Neue Materialien, Im Stadtwald, Geb. 43, 66123 Saarbruecken, Germany

SUMMARY

The use of nanoparticles in hybrid and polymeric matrices has been investigated. Different types of nanocomposites have been prepared. The ormosil type based on organoalkoxy silanes shows surprising properties with respect to solid content and relaxation. Thick films for coatings, embossed films for diffractive purposes, high temperature stable binders for glass fiber insulation materials and printing pastes have been prepared. The nanomer type with nanoparticles homogeneously dispersed in polymer matrices led to optical adhesives with improved properties with respect to T_g and thermal coefficient of expansion. The inverse nanomer type with the nanoparticles forming a percolating network was developed for photo curable optical interference layers on plastic, for example to AR coating by using simple wet chemical coating techniques.

Introduction

Hybrid materials have become a very interesting class between polymers and ceramics. Over a period of more than 20 years they have become a well established area in the field of sol-gel synthesis. This was mainly based on the properties of silicon to form stable bonds to oxygen and carbon at the same atom. Since silicon alkoxides have been used as model agents in many investigations [1-14]. Meanwhile a variety of industrial applications has been reported for example as hard coatings [15-18], dental materials [19] or low-surface free energy coatings on sanitary ware [20]. Due to the manifold possibilities to combine inorganic with organic synthetical chemistry an almost endless variability of structures exists which opens up very attractive new synthesis routes for inorganic and organic chemists. From the industrial point of view that means market and commercialization, however, very often it is difficult to find out which route will lead to useful materials for application and very often cost versus performance problems prevent the realization of interesting potentials. Whereas inorganic-organic hybrid materials based on sol-gel synthesis have been considered as more or less molecular composites, more recently the hybrid materials in combination with particulate systems have become of interest [21-33]. This systems may be called Nanomers [34) 35)] and one of the reasons for the development of this group of materials is based on the fact that inorganic nanoparticles in most cases show physical, chemical, optical or electronical properties which a single inorganic molecule

    CCC 1022-1360/00/$ 17.50+.50/0

does not show. Those properties may be semiconducting properties, non-linear optical properties or even catalytic properties. Another attractive property is the large interface of these particles which can become of importance because interfaces for example between the two phases always show properties different from the single phases. These interfacial phases may gain importance if their volume fractions become remarkable, for example as interfaces between nanoparticles and polymeric matrices. From the structural point of view, different classes of hybrid materials can be distinguished.

Classes of materials

Types of hybrids

Ormosils or <u>org</u>anically <u>mo</u>dified <u>sil</u>icates are describing a sol-gel derived material with a mainly inorganic backbone and modifying groups dispersed in this backbone. In figure 1a, a simplified structure of the type I (ormosil) is shown. In figure 1b, a more complex ormosil structure is shown, with organic chains as links between Si-atoms. This type also

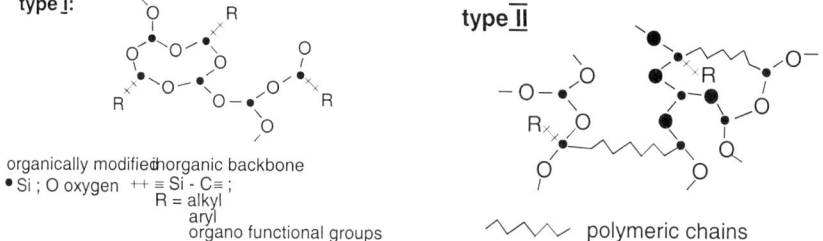

type I:

organically modified inorganic backbone
● Si ; O oxygen ++ ≡ Si - C≡ ;
R = alkyl
aryl
organo functional groups

type II

~~~ polymeric chains

Fig. 1a: Type I hybrid material: ormosils with out organic chains

Fig. 1b: Type II hybrid material ormosils with organic chains

may include interpenetrating organic chains or adsorbed organic molecules like dyes. As a type III structure, other elements as inorganic backbone former can be used, eg. ≡ Zr-O-, ≡ Ti-O-, =Al-O- or others. To link organics to the inorganic ions, complex formation (eg. acetylacetonates), acid bases or chelating ligands can be used. These types also became known as ormocers, ceramers or polycerams. If the inorganic phase forms nanoparticles (oxides, metals or semiconductors), a type IV structure is formed with a type I-III as polymer matrix. This type is called nanomer (<u>nano</u> particle containing poly<u>mer</u> type matrix), where the nanoparticle can be linked to the network or not. Figure 2 shows a structural model. For the fabrication of nanomers, different routes have been developed:

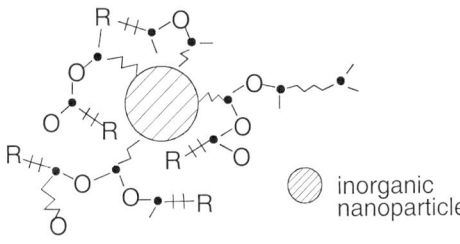

Fig. 2: Type IV hybrid material; nanomer type

The fabrication of nanoparticles in a seperate process [36)37)], based on a thermodynamical approach through a nucleation and growth process under the control of surface modifying agents. In figure 3 the nucleation frequency is shown as a function of various parameters (eq (1) and (2)), where σ in eq (2) can be controlled by surface modifiers, (e.g. β- diketones). As shown elsewhere in detail [37)], the formation of ZrO$_2$ nanoparticles can be controlled in size by the concentration of acac during hydrolysis and condensation. Due to the coverage of various amounts of acac, the ζ-potential which depends on the concentration of acid or basic surface sites, shows lower values, but as expected, the point of zero charge (p.z.c) does not change. In addition to the ζ-potential effect, the surface modification also can reduce the particle to particle interaction to the Van der Waals or weak dipole interaction and thus prevents the formation of hard agglomerates. Weakly agglomerated surface modified powders have been obtained from many systems (figure 4), fully redispersible due to the appropriate surface modification [36)38-40)].

$$x\, Z(RO)_4 + y\, H_2O \longrightarrow x\, ZrO_2 + y - x\, H_2O + x\, HOR$$

$$I = A \cdot e^{\dfrac{-(\triangle G_i + \triangle G_D)}{kT}} \qquad (1)$$

$$A = 2 n_v \cdot \upsilon^{1/3} \, \frac{kT}{h} \, \sqrt{\frac{\sigma}{kT}} \qquad (2)$$

$\triangle G_i$: Gibbs free energy for nucleation
$\triangle G_D$: Gibbs free energy for diffusion
n_v : number of nucleating or crystallizing species per unit volume of the liquid
υ : volume per formula unit of the nucleating species
σ : interfacial free energy

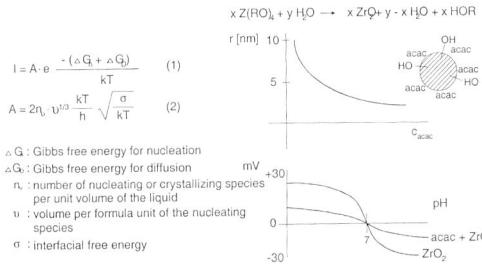

Fig. 3: Dependencies of the nucleation frequency

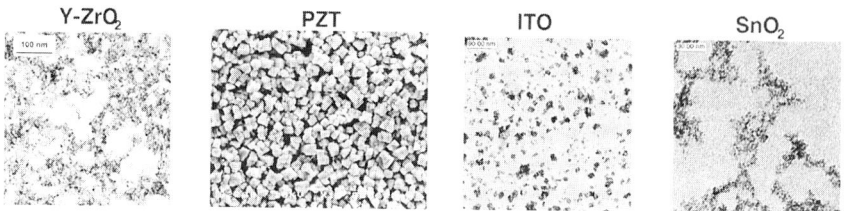

Fig. 4: Examples of agglomerate-free nanoparticles (after [36])

Ormosils

One of the simplest examples for ormosils are hybrids fabricated from tetraethylorthosilicate and alkylalkoxysilanes. This type of ormosils have been used as so called spin-on glasses in microelectronics [40] or as mild abrasives for skin treatment [41]. Since the alkylsilanes are reducing the network-connectivity and as a consequence the brittleness of the spin-on glasses also will be reduced, liquids are obtained which can be used in coating processes like spin- and dip coating. However, the film thickness which can be obtained by this method is rather limited. In a spin or dip coating process one or two μm can not be exceeded. This phenomenon has been investigated from Fred Lange and co-workers which found out that in given sol-gel systems, the strength of the material, the shrinkage during drying and firing, the solid content and the relaxation behavior are the most important parameters for the obtainable film thickness. They concluded that in sol-gel materials thickness above 1 μm are difficult to obtain. Since the extend of shrinkage is directly connected with solid content of a dried film, the solid content in gels in general is far away from the theoretical package density due to the increasing brittleness of the inorganic network during condensation which leaves pores and voids in the structure. It seemed to be of interest how far this draw back could be overcome by substituting a part of a sol-gel-derived SiO_2 from hydrolysis and condensation of TEOS (tetraethylorthosilicate) and MTEOS (methyltrietoxysilane) by already condensed SiO_2 in form of nanoparticles. For this reason, colloidal silica has been used in form of aqueous suspension (Bayer chemical company) and reacted with TEOS and MTOS or methyltrimethoxysilane (MTMEOS) respectively. The experimental process is described in detail elsewhere [42,43]. As a measure for the influence of SiO_2 nanoparticles, the critical thickness was measured and defined as a thickness of a film obtained by a one step dip-coating process after firing at 500 °C on silica substrates without showing a single crack. In the dependence of the critical thickness as a function of the SiO_2-content is shown. As one clearly can see, the critical thickness increases strongly with the increasing SiO_2-content and reaches a maximum at about 40 % with values around 14 μm (figure 5). The MTMEOS system shows a little bit higher values than MTEOS. If the solid content exceeds 40 wt.-%, the critical thickness decreases which can be interpreted with the fact that the relative concentration of methyl groups in the system is reduced more and more leading to an increased particle interaction approaching typical inorganic sol-gel coating thicknesses. It was of interest to investigate whether there

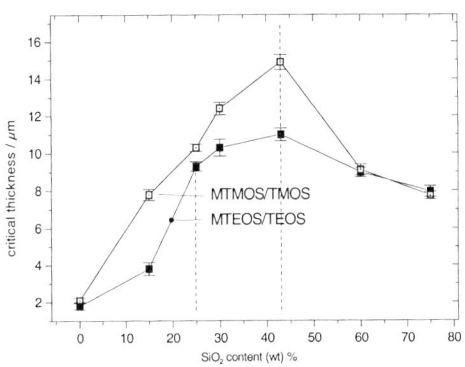

Fig. 5: Critical thickness as a function of the SiO$_2$-content in different ormosils

are other effects of the SiO$_2$ nanoparticles on the condensation kinetics of the system. For this reason, a T- andQ-Ana lysis were carried out in the SiO$_2$-containing and in the SiO$_2$-free system. In figure 6a and b the T analysis is given. (T and Q analysis show the connectivity of condensed silans: T^1 means a Si-R con-

nected by one oxygen bridge to another Si:R-Si-O-Si≡, T^2 means two bridges an so on. Q^1 means ≥Si-O-Si≥ , Q^2 =Si$^{-O-Si\equiv}_{-O-Si\equiv}$ and so on.) The T and Q values are measured by ^{29}Si-NMR spectroscopy. Figure 6 clearly shows that the SiO$_2$-nanoparticle containing systems lead to lower T^2 and higher T^3 values after the same reaction times. For Q^3 and Q^4 a similar

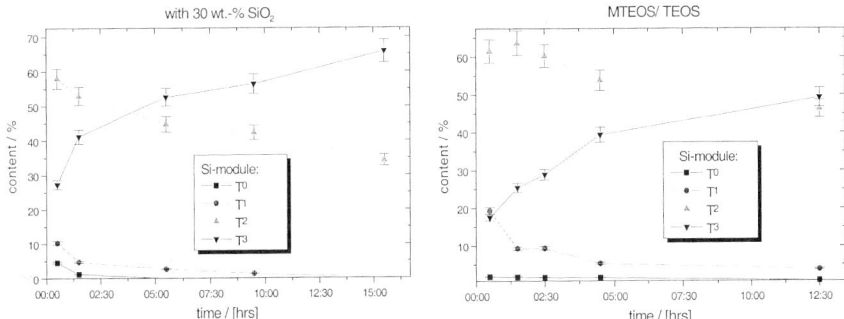

Fig. 6a: T^0- T^3 measurements for MTEOS/TEOS with SiO$_2$

Fig. 6b: MTEOS/TEOS without SiO$_2$

behavior is observed. However, the increased condensation kinetics does not explain the higher critical thickness. The particle size analysis as a function of time (ageing) shows that in the SiO$_2$-sol containing systems the particle size increase is much faster than in the

SiO$_2$-sol free systems (3-400 nm versus 6-800 nm). As photon correlation spectroscopy measurements as a function of the aging time show, this can be interpreted by a type of "nucleation effect", with the hybrid sol forming a gel-like structure around the SiO$_2$ nanoparticles. It is anticipated that a core-shell structure is formed as schematically shown in figure 7. As shown in [44], during drying of hybrid gels the formation of hydrophobic

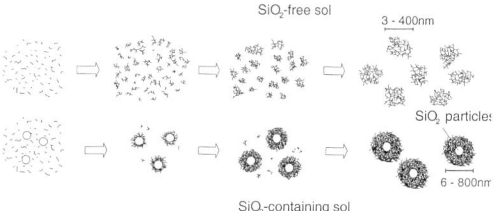

pores takes place, measured by absorption isotherms with water vapor. This leads to the conclusion that methyl-groups are preferringly present in interfaces thus reducing the particle to particle interaction, which otherwise would be very strong by the formation of

Fig. 7: Structure formation in the silica particle containing MTEOS/TEOS systems; particle sizes measured by PCS.

Si-O-Si or hydrogen bonds. This structure should lead to an enhanced relaxation. Following the hypothesis of an enhanced flexibility as an important reason for the increased critical thickness, the burn out temperatures of CH$_3$ was investigated by TGA and DTA measurements. The oxidation peaks maximum of methyl groups and ethoxy groups have been evaluated and the maximum temperatures are shown in figure 8. As one clearly can see, as a function of the SiO$_2$-content, the oxidation peak of the methyl-group is shifted

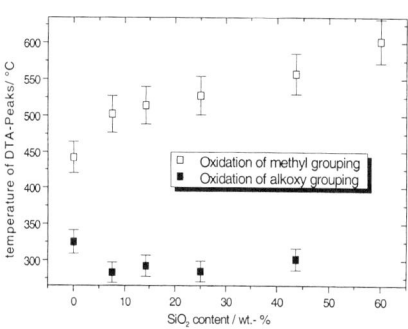

Fig. 8: Oxidation peaks maximum of methyl groups and ethoxy groups

from 440 °C to 600 °C. This means that by establishing the appropriate SiO$_2$-content a temperature can be chosen where the methyl-groups disappear. In figure 9 a-b the corresponding IR-spectra are shown, also clearly depicting the higher temperature stability of SiO$_2$-sol containing systems (peaks at the 2900 and 1250 cm^{-1} regium). These results support the hypothesis of an increased relaxation at higher temperatures as an important factor for an increased critical thickness. To investigate the effect of these coatings on glass fibers, a fiber sample coated with the SiO$_2$- containing sol was investigated in a fiber elongation viscosimeter (figure 10). The results show that up to 1000°C no viscous flow is detectable. Moreover, above T$_g$

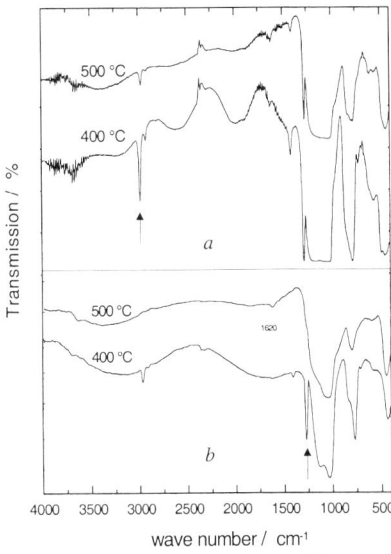

Fig. 9a: MTEOS/TEOS / 25% SiO$_2$,
Fig. 9b: MTEOS/TEOS without SiO$_2$

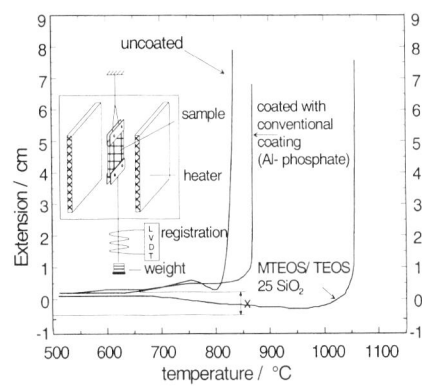

Fig. 10: Fiber elongation viscosimetry of uncoated Al-phosphate and SiO$_2$ sol coated fibres

of the basic glass a contraction (x in figure 10) is observed, which is attributed to the sintering process of the coating. This means, that a densification takes place without a viscous flow mechanism. Otherwise a contraction against the elongation force of the viscosimeter should not be possible. Summarizing one can say that based on a structure model, a highly relaxable sol-gel coating system could be developed which leads to a substantial increase of the critical thickness compared to conventional sol-gel systems. This type of ormosils can be considered as ormosil nanocomposites. The sintering and densification of a gel with a composition of SiO$_2$ glass without viscous flow opens up the possibility to use these materials as inorganic high temperature resistant binders with the advantage not to change the shape pf the part by viscous flow. As shown by Mennig et. al, gel layers can be embossed at room temperature and then transferred into SiO$_2$ glass without changing the shape of the patterns [45]. As recently shown by Kalleder et al. [46], the system has been used as binder for fine-line printing processes with Ag as conductive agent. By mixing silver powder with only about 2% of the described SiO$_2$-containing binder and screen printing additives, a printing paste could be synthesized having an viscosity appropriate for fine-line printing. Conventional conductive printing pastes are based on glass frits as a viscous flow binder which only can be densified above the T$_g$ of the glass. This leads to a viscous flow and as a consequence in restrictions in the line shape (rounding) and in the line distances (broadening). This does not take place with the new nano-based binder ormosil, since viscous flow sintering is avoided. The paste was prepared by using the SiO$_2$ containing sol and by mixing it with Ag

(30μm), tetraethylene glycol and hydroxypropylcellulose [47] to obtain the required rheology. After printing the lines were heated to 570 °C for 20 min without a broadening to be observed. In figure 11, a printed pattern is shown. No uncontrolled line broadening takes place. Other interesting applications are embossing techniques. The ormosil nanocomposite

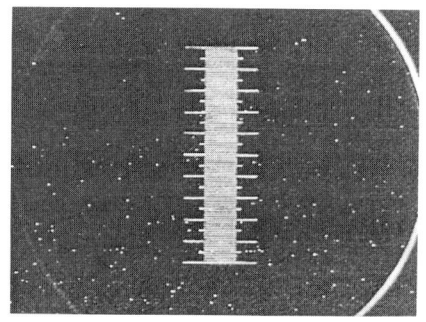

film can be easily embossed after coating since due to the reduced particle-to-particle interaction the film is very soft (soft gel film, SGF) (figure 12). Silicone rubber stampers can be used to emboss these systems. Due to the very soft gels, silicone rubber stampers or rolls can be used, making this process cheap. In figure 13, embossed light traps on a glass

Fig. 11: a 1mm in length grid line

substrate for the efficiency increase of solar collectors and a hologram embossed on a metal substrate are shown. The films not only can be prepared from SiO$_2$, but also from TiO$_2$ nanoparticles to produce high refractive index patterns. The 120 °C cured coating

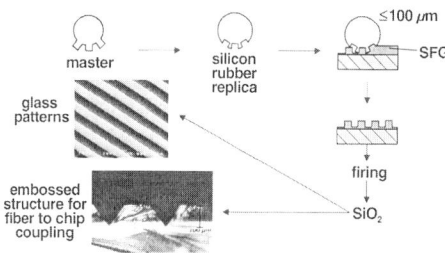

Fig. 12: Principles of the process for making micropatterns using ormosil nanocomposites (SGF-process)

Fig. 13: Glass light traps on solar collectors

systems show high flexibility compared to inorganic sol-gel coatings and, due to the reduced burn out kinetics of the methyls, this elasticity is maintained up to higher temperatures. This led to the idea to investigate how far this system can be used as a new binder for glass fiber mats. For this reason, the binder was sprayed on an uncoated glass fiber and the elasticity was measured. In figure 14 the result is shown. The glass fiber mat behaves the same way as conventional phenolic resin bonded glass fibers (figure 14). This results have been used to build up an industrial process for the fabrication of ormosil bonded glass fibers. The organic content of these glass fibers is only in the range of 2%

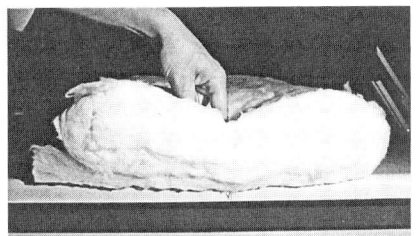

Fig. 14: "White" glass fiber mats showing very elastic behavior

Dämmstoffe Company, Germany.

which means that in the case of fire no poisonous crack products are produced. The glass fiber mat withstands temperatures up to 600 °C without losing their insulating properties. It can be easily recycled and used as fire insulation materials. The process was commercialized with Pfleiderer

Nanomers

The dispersion of nanoparticles in more polymer-like matrices leads to the concept of nanomers (nanoparticle reinforced polymers). As already shown elsewhere [47], a SiO_2 particle reinforced adhesive for fiber to chip coupling has been developed by use of a high degree of filling of epoxides. Detailed investigations about the incorporation of the SiO_2 nanoparticles have shown that it is of great importance for obtaining a homogenous dispersion to control the surface free energy of these particles in order to adapt them to the polymeric matrix system. As polymeric matrix system an uncured epoxy-system and as fillers 6 nm SiO_2 from commercially available sols have been used. The idea was to investigate how far the polymer properties with respect to the cte (coefficient of thermal expansion), modulus and T_g may be affected advantageously without disturbing application properties like viscosity. It has to be kept below 15 mPas·s, since the adhesive has to be filled into a narrow slit of 3 μm (figure 15). For this reason the viscosity has been

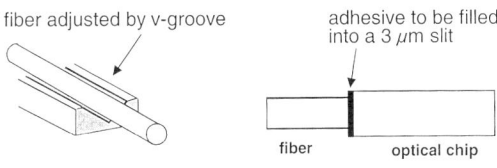

fiber adjusted by v-groove

adhesive to be filled into a 3 μm slit

fiber   optical chip

Fig. 15: Principle of fiber to chip coupling by adhesive bounding

investigated as a function of the silica content. From interfacial thermodynamical considerations it was clear that the silica surface has to be modified in order to reduce the interfacial free energy because the hydrophilic OH-group containing silica and the relatively hydrophilic epoxy monomer. On the other hand the surface modification should not contribute substantially to the volume in order not to change the properties of the polymerized system. For this reason tetramethylammoniumhydroxid (TMAH) has been chosen, since, due to the polarity of the surface of the SiO_2 and the molecule a reasonable interaction between the two components was anticipated. The TMAH would not be a perfect surface modifier in aqueous systems but it was expected to be so in non-aqueous systems. In figure 16 the influence of surface modification and SiO_2-content on the viscosity is shown. As one

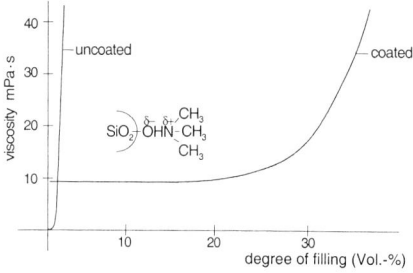

Fig. 16: Viscosity of the uncured epoxy as a function of the SiO_2 (uncoated and coated with TMAH)

clearly can see, in the unmodified case the viscosity increases drastically with very low SiO_2-content leading to not usable systems. After the surface modification with TMAH the viscosity remains stable up to 20 or 25 vol.-% (representing about 40 weight-%) of 6 nm SiO_2. The properties of the polymers change remarkable. In figure 17, the TEM micrograph of a sytem filled with about 25 vol-% of SiO_2 is shown. It clearly depicts the homogeneous densification of SiO_2. In figure 18 the modulus as a function temperature and SiO_2-content is shown. With increasing

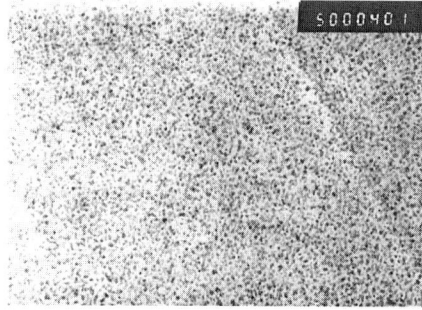

Fig. 17: TEM micrograph

SiO_2-content a rubbery plateau is formed above T_g. The cte also is changing remarkably. (From 200 to $150 \cdot 10^{-6}$/K in the 120 °C regime and from 110 to $80 \cdot 10^{-6}$/K in the 50 °C regime). The new nanocomposite adhesive shows several advantages: lower shrinkage, no part curing, faster curing and is commercialized already [48]. A completely different type of nanomers is accessible by "inversal structure", meaning that not the polymer forms the percolating phase but the nanoparticles. The details are described elsewhere [48]. In this case TiO_2 and SiO_2 nanoparticles prepared by a sol-gel route have been modified with glyceroloxypropyl triethoxysilane (TiO_2) and methacryloxypropyl

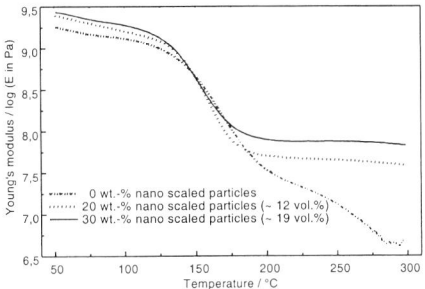

Fig. 18: Young modulus vs. temperature for SiO₂ containing nanocomposites

triethoxysilane (SiO₂), respectively. If these nanoparticle sols are used for coatings on polymers, for example by a dip coating process, they can be photo-polymerized if the appropriate photo-catalysts are added (radical photo-initiators for the methacrylates and ionic photo-initiators for epoxides) [49]. Measuring the density of these materials by various methods leads to the surprising effect that the package density is much higher as expected, for example in the case of SiO₂ refractive indices above 1.9 have been obtained suggesting a package density in the range of 90% of the theoretical density. In the schematic process of the layer formation is given (figure 19). Using this technique, interference layers (from SiO₂ and TiO₂) have been fabricated on plastics, only using a wet coating process in combination with photocuring. In figure 20 an interference filter is shown, produced by dip coating and angle dependant dip coating. Summarizing it is to say that this technique leads to a completely new class of nanomers that means polymers which do not have a polymeric chain but nanoparticles which are connected by short organic links which makes them flexible enough to be used on polymers which show an extremely high package density. More investigations of this type of materials is necessary in order to learn more about structure and properties [49].

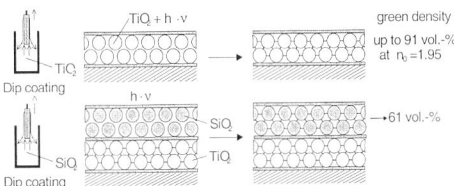

Fig. 19: Interference filters produced by the universal nanomer process on polycarbonate

54

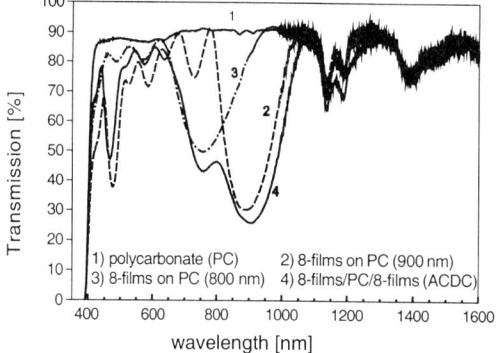

Fig. 20: Scheme of the process of the formation of microsol nanomer layers

Conclusions

The combination of polymer type of materials with inorganic components can lead to interesting materials with a high potential for industrial applications. A different type of structures can be synthesized by using various inorganic or organic precursors and by using appropriate processing techniques. This is a very important factor since the properties depend strongly on these structures. And moreover the introduction of nanoparticles leads to additional tools for material tailors. First by the functional properties of the nanoparticles for example by effecting the reaction kinetics and second by the generation of an interfacial face to the polymer which is different in structure from the environment and which leads to a change of the polymer properties. In addition to this the nanoparticles may lead to so called inversed structures, a class of materials which has not been investigated in detail so far.

Acknowledgment

The authors greatly acknowledge the financial support of the industry, the State of the Saarland for the work.

References:

1. H. Krug, H. Schmidt, *Proc. of Eur. Wsh. Hyb. Org.-Inorg. Mat. Syn., Prop.,* Appl. **1**, 127-141, (1993)
2. R. Kasemann, H. Schmidt, *Proc. of Eur. Wsh. Hyb. Org.-Inorg. Mat. Syn., Prop., Appl.* **1**, 171-180, (1993)
3. D. Hoebbel, T. Reinert, K. Endres, H. Schmidt, A. Kayan, E. Arpac, *Proc. of Eur. Wsh. Hyb. Org.-Inorg. Mat. Syn., Prop.,* Appl. **1**, 319-323, (1993)
4. H. Schmidt, P. W. Oliveira, H. Krug, *MRS Symposia Proceedings,* **435**, 13-14, (1996)
5. V. Gerhard, H. Schmidt; U. Dreier, *MRS Symposia Proceedings,* **435**, 455-460, (1996)
6. P. Müller, C. Becker, H. Schmidt, *MRS Symposia Proceedings,* **519**, 387-393, (1996)
7. H. Schmidt, E. Arpac, H. Schirra, S. Sepeur, G. Jonschker, *MRS Symposia Proceedings,* **519**, 297-308, (1998)
8. H. Schmidt, E. Geiter, M. Mennig, H. Krug, C. Becker, R.-P. Winkler, *J. Sol-Gel Sci. Tech.* **13**, 397-404, (1998)
9. M. Mennig, M. Zahnhausen, H. Schmidt, *Proc. of SPIE,* **3469**, 68-78, (1998)
10. C. Becker, P. Müller, H. Schmidt, *Proc. of SPIE* **3469**, 88-98, (1998)
11. H.K. Schmid, M. Aslan, S. Assmann, R. Nass, H. Schmidt, *J. Eur. Cer. Soc.* **18**, 1, 39-49, (1998)
12. H.K. Schmid, *Proc. of EUROMAT 5*, 5-10, (1997)

13. H. Schmidt, H. Krug, B. Zeitz, E. Geiter, *Proc. of SPIE* **3136**, 220-228, (1997)
14. P. Judeinstein, P.W. Oliveira, H. Krug, H. Schmidt, *Adv. Mat. Op. Elec.* **7**, 123-133, (1997)
15. G.W. Wagner, S. Sepeur, R. Kasemann, H. Schmidt, *Proc. of Sol-Gel Production*, 193-198, (1998)
16. H. Schmidt, B. Seiferling, G. Philipp, K. Deichmann, *Proc of Ultrastr. Proces. Cer., Glas. Comp., Wiley*, 651-660, (1988)
17. R. Kasemann, H. Schmidt, *N. J. Chem.* **18**, 10, 1117-1123, (1994)
18. S. Langenfeld, G. Jonschker, H. Schmidt, *Mat. Wiss. Werkst.* **28**, 1-15, (1997)
19. H. Wolter et al. DE-PS 4133494 C2 (28.3.1996)
20. Technical information about "Wondergliss", Duravit, (1999)
21. J. E. Mark, *Proc. of Int. SAMPE Tech. Conf.*, **27**, 539-548, (1995)
22. T. Saegusa, *Proc. of Macromol. Symp.*, 719-29, (1995)
23. B. M. Novak, M. W. Ellsworth, C. Verrier, *Polym. Mater. Sci. Eng.,* **70**, 266-7, (1993)
24. C. L. Beaudry, L. C. Klein, *Polym. Mater. Sci. Eng.*, **73**, 431-2, (1995)
25. C. L. Beaudry, L. C., Klein, *Proc. of ACS Symp. Ser.*, **622**, 382-94, (1996)
26. J. Wen, G. L. Wilkes, *Chem. Mater.*, **8**(8), 1667-1681, (1996)
27. P. B. Leezenberg, M. D. Fayer, C. W. Frank, *Pure Appl. Chem.*, **68**(7), 1381-1388, (1996)
28. Z. Gao, Z. Zhao, Y. Ou, Z. Qi, F. Wang, *Polym. Int.*, **40**(3), 187-192, (1996)
29. M. I. Sarwar, Z. Ahmad, **Adv. Mater.**, 73-77, (1997)
30. M. Motomatsu, T. Takahashi, H.-Y. Nie, W. Mizutani, H. Tokumoto, *Polymer*, **38**(1), 177-182, (1997)
31. J. F. Gerard, H. Kaddami, J. P. Pascault, *Proc. of Eurofillers 97*, 407-410, (1997)
32. A. B. Wojcik, A. Ting, L. C. Klein, *Mater. Sci. Eng.*, C, **C6**(2,3), 115-120, (1998)
33. E. Reynaud, C. Gauthier, J. Perez, *Rev. Metall./Cah. Inf. Tech.*, **96**(2), 169-176, (1999)
34. B. Braune, P. Müller, H. Schmidt, *Proc. of SPIE*, **3469**, 124-132, (1998)
35. *Chemistry and technology of nanoparticles: preparation, processing and application.* Proc. of EUROMAT, 5-10, (1997)
36. H. Schmidt, R. Nonninger, *Proc. of Fine, Ultrafine and Nano Powders '98*, (1998)
37. H. Schmidt, *Kona powder and particle (14)*, 92-103, (1996)
38. R. Nass, S. Albayrak, M. Aslan, H. Schmidt, *Proc. of CIMTEC*, **11**, 47-54, (1995)
39. R. Nass, S. Albayrak, M. Aslan, H. Schmidt, *ACerS*, **51**, 591-595, (1995)
40. A. M. Ron, M. Fibich, *Sol. St. Commun.*, **58**(12), 869-72, (1986)
41. H. Schmidt et al. *Abrasiv wirkende Mittel und deren Verwendung* DE-PS 3048369 C2 (5.1.1983)
42. G. Jonschker, *Ph. D. Thesis*, University of Saarland, Saarbrucken, Germany (1998)
43. H. Schmidt, G. Jonschker, S. Goedicke, M. Mennig, *Proc. of 10th Int. Ws. on Glas., Cer., Hyb. Nanocomp.*, in print (1999)
44. H. Schmidt, H. Böttner, *Iler 91-Symposia, Advances in Chem. Ser.* 234 ACS, 419-432, (1994)
45 M. Mennig, G. Jonschker, H. Schmidt, *Proc. of SPIE*, **1758**, 125-134, (1992)
46. A. Kalleder, R. Kreutzer, M. Mennig, H. Schmidt: *Proc. of Euromat 99*, in print
47. Th. Koch, M. Mennig, H. Schmidt, *Proc. of CIMTEC*, **17**, 681-688, (1999)
48. H. Krug et al, *Nanostrukturierte Formkörper und Schichten sowie Verfahren zu deren Herstellung.* DE-OS, 19719948 A1 (19.11.1998)
49. M. Mennig, P.W. Oliveira, A. Frantzen, H. Schmidt, *Proc of 2nd ICCG*, **351**, 225-229, (1998)

"What is KRI?"

- R&D based on excellent tradition and innovation -

Masaaki Takeuchi

Kansai Research Institute, Inc.
Kyoto Research Park, 17,Chudoji, Minami-machi, Shimogyo-ku, Kyoto 600-8813,
Japan

SUMMARY: As far as KRI's clients were concerned, KRI's appeal was not
only in having researchers who had a high level of expert knowledge, but it
was also in the efficiency with which high risk jobs were conducted, and the
speed of response time, especially when corporations made comparisons with
their own in-house labs. KRI, while being a facility for contract research, is
also capable of becoming a profitable corporation, mostly thanks to the
adoption of this system. KRI sincerely hopes to continue serving as a group
made up of quality professionals who embrace a dream, and to contribute to
society through its achievements. At the same time, the company strives to be
sensitive to changing needs, and, of course, to conduct R&D and consulting
activities which will ensure success for its clients.

1. What is KRI

1.1 The First Comprehensive Think Tank Combining a Consulting Division and a Contract Research Division

KRI, an abbreviation for Kansai Research Institute, Inc., was established in 1987 under the
auspices of OG (Osaka Gas), which has a 100% investment in it. It was established as
Japan's first comprehensive think tank which has both a consulting division and a contract
research division. Differing from the Japanese style of management seen so far, this think
tank with its own unique style of management achieved recognition for its existence and
capabilities.

1.2 KRI's Current Areas of Business

The revenue for 1998 reached approximately 4.3 billion yen and the company grew to over
300 people, but KRI has not yet broken through the boundary of small to medium size

companies. However, in the industrial arenas of R&D-related business, KRI is the leading company with no competitors yet making an appearance. The Consulting Division, the Contract Research Division, and the Technology Trend Analysis Division were established in 1987, and in April, 1999, the Analysis Center was set up. KRI is a subsidiary with 100% investment from OG, but since its establishment, to the best of its abilities, KRI has kept a check on the loaning of personnel from OG. The KRI we see today was shaped in the climate created out of being a venture company.

2. The Circumstances of KRI's Establishment

2.1 Why Was KRI Established?

At that time in Japan, there were no genuine think tanks that did consulting and had laboratories, and all the major corporations in Japan had to use think tanks outside of Japan, spending a lot of money on R&D for new fields. Therefore, Osaka Gas believed that a genuine think tank could be successfully established in Japan. A business tie-up with SRI was realized.

2.2 About KRI's Ideas on Management

From the beginning, we have intended for KRI to become a genuine think tank which combines a consulting division with a contract research division, and after achieving the successful results, we have sought to gain the trust of our clients, which would further contribute to their development. We are also aiming to become an ideal think tank whereby researchers and consultants can exhibit their abilities to the fullest.

2.3 The Way of Thinking about Outsourcing in Japan at the Time of KRI's Establishment

An awareness of outsourcing had begun to take hold in Japan, but for corporations within Japan, the field of R&D was still kept strictly in-house at the time of KRI's establishment, and it was a time when the environment for outsourcing R&D was still quite harsh.

2.4 The Establishment of the Japan Research Industries Association and the Trends of Research and Industry in Japan

During the period when private research facilities were gradually playing a bigger role in Japan, efforts were being made to promote the smooth development of both research and activities supporting research. To this end, the Japan Research Industries Association was established in March of 1991 with the objective of contributing to the industrial development of Japan. Now, the official role of contract research facilities in Japan was at last clearly confirmed. KRI immediately began to participate. However, the fact that there were only 5 companies which were registered as independent R&D corporations, is demonstrating the difficulty of establishing companies based on R&D-related business.

2.5 Extending the Scope of Outsourcing and R&D Outsourcing

Around the time of 1991, Japan was going through a recession which meant that companies had to take sole responsibility for their business, making proceeding with R&D more difficult. Yet, at the same time, companies preferred to outsource their R&D to speed up the time it took to get to see the results and to avoid risk besides taking care of their core technologies. Moreover, the fact that restructuring had created a shortage of highly skilled researchers increased the value of contract research facilities like KRI which had maintained a high technical level.

3. Concerning the Steps Required to Begin Contract Work at KRI

3.1 A shift to outsourcing on the project- proposal model /strategic model

KRI was not such a strong company at its inception, and most of its business was from the promotion of projects partially contracted by clients. However, from an expansion of KRI's scale and growth in strength as a company, KRI came to develop projects based on its own original ideas which were used in presentations. This fostered KRI's abilities to become a company capable of its own business undertakings and led to a shift to contracting projects based on the project proposal model and the strategic model. Presently, these types account for approximately 70% of all projects contracted.

3.2   Steps for Acquiring New Clients

The abilities of KRI members to make proposals, develop, and conceptualize all with an eye on the times, in the climate at KRI whereby specialists of varying backgrounds and fields coexist, are born out of a stimulating environment.   KRI's biggest strength is not in gathering together a group of experts all of the same quality.   Rather, KRI's biggest strengths lie in having a group of experts of different qualities who can value each other's techniques from different perspectives as well as in fostering a climate which sublimates new ideas. There are still no other corporations in Japan with such strengths.   Each of the ideas of KRI members is represented as a think piece put on one piece of paper (A-4 size) and is used as a sales tool.   Projects based on think piece facilitate meetings between KRI and its clients. The industrial property rights of any project results are basically belong to the clients. However, the proportion of property rights granted to KRI, and to the client will depend on the monetary amount of the contract.   This is in order to avoid problems after an agreement has been reached at the contracting stage.

4. Concerning the Corporate Activities of KRI to Date

In the 1st period   ('86-'94), KRI was in a state of lateral growth in its actual profits and number of personnel, as it experienced a difficult period of management and an unclear future.   In the 2nd period ('95-'98), despite the recession, the environment and times improved for contract research companies, and efforts were made both to be more aggressive in marketing practices and to increase the scope of business.   From there, KRI was able to achieve a yearly rapid growth of more than 30% in the last 5 years.   The number of contracted projects almost doubled from 320 in 1995 to 620   in 1998.   Several factors which were important to the development of research are mentioned next.   One was the feeling on the part of researchers and consultants that their research themes challenged them. Another was that they had a sense or feeling of the approach they took toward their research and the results they got from it. And one more was the importance of public support for the research themes that had been chosen.   KRI didn't concern itself solely with big revolutions in technology based on "Something New".   Besides that, KRI's important accomplishments worth mentioning include such things as tackling with equal energy and zeal areas in technology that deal with improving the usability of everyday goods and also responding diligently to the needs of the client.   Moreover, whatever project formation was considered most suitable in order to get results quickly, for the sake of that, KRI would coordinate matters so innovations of the organization were facilitated, and the abilities of researchers and consultants could be fully brought out.   Domestically, without taking OG into account, the percentage of clients in the Kansai area is lower than that of the Kanto area.   Because of this, marketing efforts organized around the Kanto area are directed by the Tokyo Market Development Division at potential clients.   On the one hand, foreign clients amount to only 10% of the total, but with the economic recovery of South Korea, and aggressive marketing in North America and Europe, it is expected that this percentage will increase.

5. KRI's Research Administration

5.1   Personnel Policies and Methods of Evaluation

Due to the differences in scope, one would hesitate to make comparisons, but KRI was originally established with the intention of being a Japanese version of SRI, and as such it incorporated SRI's management administration.   Therefore, KRI's system of management administration, which basically emphasized the importance of SRI's efficiency in getting results, was most favored by KRI's clients.   As far as KRI's clients were concerned, KRI's appeal was not only in having researchers who had a high level of expert knowledge, but it was also in the efficiency with which high risk jobs were conducted, and the speed of response time, especially when corporations made comparisons with their own in-house labs. KRI, while being a facility for contract research, is also capable of becoming a profitable corporation, mostly thanks to the adoption of this system.   From the beginning, personnel policies and evaluation methods for all employees followed those of SRI, with yearly contracts, no seniority system, no discrimination based on gender or nationality, and all members evaluated on their own merits alone.   The majority of the research staff at KRI supports this system.

6. The Corporate Climate at KRI

6.1 Creating A Body of Researchers Who Want to Continue Doing the Work They Like

As has been previously stated, the main source of research contracts secured by KRI is not from projects commissioned by corporations. Rather, the main source comes from proposal-based research coming out of contracts secured after aggressive presentations by KRI targeted at specific corporations. As research following this pattern advances, the number of people changing jobs will increase. Examples of such cases include, members of large corporations who are no longer able to continue research due to restructuring despite single-mindedly conducting their own research; people who prefer to do laboratory research as scientists over obtaining a managerial position; people at the level of assistant or assistant professor within the university system who wish to conduct their research more freely. It is taken for granted that what is sought is not the desire to do research one likes as one likes, but the ability to produce results under the given time constraints.

6.2 The Climate Surrounding Career Path Choices

In this way, KRI fosters a climate which encourages the development of technologies and the intellectual stimulation in meetings among researchers of different fields; all of which cannot be gained at a single corporation or research laboratory. KRI offers a climate whereby employees with KRI as their base can pursue their own career aspirations, while contributing to KRI by way of their own achievements. This is not to say that KRI urges employees to leave the company quickly. Rather, what it meant is that while KRI as a workplace retains its appeal, it is hoped that employees will continue to strive hard at their research. Furthermore, KRI by no means wants employees to feel tied to the company. As part of choosing one's career path, KRI wants each individual to make his or her own decisions independently.

7. The role of the International Advisory Council

7.1 The function of the external management board

Since the very beginning of its establishment, when KRI had nearly 30 employees, the International Advisory Council was organized. The board consisted of many internationally well-known people. Though this first seemed too much for such a young company, being backed up by the board is now considered to have been an appropriate decision. During the period when KRI was being established, the names of the members contributed greatly to KRI, lending support as KRI started marketing itself abroad before it had time to demonstrate its own reputation and credibility.

8. An introduction to the technologies and topics owned by KRI

8.1 An overall image of KRI's technologies

KRI has been producing research results of quality based on its accumulated wisdom and ideas grown in a unique climate. In the next decade, the main current in technology will be related to the environment, energy, and electronics. Each research center at KRI has been strategically acquiring technological know-how and capabilities relevant to their particular field. Today, I would like to briefly introduce KRI, concentrating on its three research centers, the Advanced Materials Research Center, the Surface Science Research Center and the SQUID Laboratory. Some examples will follow with a short explanation.

8.2 Advanced Materials Research Center

This center consists of six laboratories. Among these six, the Inorganic Fine Lab. has been active in conducting studies based on the sol-gel method and organic inorganic molecular hybrid technologies. Another major lab, the Polymer Lab. has been trying to find functions and characteristics of novel polymer materials based upon the unique polymer synthetic technologies.

(1) Introduction of the plans for Sol-Gel technology at KRI:
The Sol-Gel method is one of KRI's important technologies, and is similar to organic polymer synthetic technology though it is an inorganic synthetic technology. Organic

polymer synthetic technology is representative of projects undertaken at KRI, and KRI is the best company for this technology.

The following methods are now reviewing at KRI.

    a.   The Micro-Capsule Sol-Gel Method of which various functional chemical compounds are encapsulated with inorganic chemical compounds.

    b.   The Emulsion Sol-Gel Method which is a result of a combination of a surfactant and sol-gel method.

    c.   The Combinatorial Sol-Gel Method of which combinatorial chemicals are useful in novel materials research.

(2)   Novel Elemental Technology for Plastics Materials Recycling:

At KRI, accumulated polymer technology is applied to waste plastics recycling. This novel technology which solves contamination problems leading to the deterioration as materials, has been developed and published. Concretely speaking, KRI succeeded in developing a new polymer called a premending reagent which prevents performance decrease and encapsulates contamination included in waste plastics without removing them.

8.3   Surface Science Research Center

Material surfaces and interfaces play an important role in the development of intelligent materials with applications in self-regulation, self-response, self-decomposition, self-diagnosis and self-identification among others. The Surface Science Research Center develops intelligent materials and related applications technology by focusing on these physical and chemical phenomena.

(3)   Determination of catalyst structure by Atomic Force Microscope technique:

This latest technological development enables us to see ultra-fine particles directly at nano-order level, however, to do so, a special instrument and an excellent technique are both indispensable.

Generally speaking, it has been fairly difficult to take a clear picture at a nano-order level even when using Atomic Force Microscope Technology.

8.4 SQUID Lab.

SQUID Laboratory of which KRI is proud, both experienced and is responsible for the first technology advances in unique and superconductive sensing technology by using highly sensitive magnetic sensors. The lab is directing quite promising technology which as successfully been used in the development of new equipment.

(4) Development of superconductivity/magnet sensing technology

The most sensitive magnetic sensors the world has ever seen, SQUIDs detect ultralow magnetic fields - as low as one billionth of the Earth's magnetic field (0.4 gauss) -not detectable with conventional devices and so promise new possibilities and a broad field of applications [medicine (magnetoencephalography and agnetocardiography), nondestructive material inspection and geophysical exploration].

9. Strategic development of the materials industry ('60s to present; looking toward the 21st century)

Many companies in the materials industry pursued improvements in productivity in the '60s. The period from 1970 to the mid '80s saw a shift to new business management based on strategies aimed at aggressive diversification. In late '80s, strategies for the kind of diversification and business expansion which would distinguish a materials company from others were devised. At present, during a long period of inactivity, even more drastic restructuring strategies are to be carried out.

One question we are faced with as we enter the 21st century: Will materials producers be able to maintain business growth and continue to survive only depending on their principal business based upon competing with other producers in the same market while they slow down the development of new business? In the new era, it is said that materials producers must be creative more than productive. Developing new business is the premise for growth in the future. Success is dependent upon innovation in research development and technology management. Advantages lie in quality not quantity in terms of the way we respond to advanced society of the future. Rapid response is indispensable especially in changing and developing fields. These are reasons that a company outsources R&D. It is needless to say which is more important, to be the champion of quantity, or to be the finest in quality.

10.   The Role of New Business and Research Development

The market for most company's core businesses is now ripe.   Companies tend to compete for a limited slice of the market.   New business development must focus on specializing and the high functionality of commodities.   Niche markets are also required in order to acquire a complete line of products.   Growth markets depend on and are led by consumers needs and changes in lifestyles and industry trends such as in the areas of information/communication (multimedia), health (health care), and the environment.   Leaders of the markets are required to have concepts that are needs-oriented rather than seeds-oriented.

11.   Company Objectives of KRI

KRI is a contract research company which conducts R&D for leading and advanced science/technologies.   The range of business reaches from basic research and applied research, to the engineering of proto-type manufacturing.   The results of information and studies produced by KRI represent active innovation processes.   In R&D, a coordinating role is very important to draw out the capabilities of researchers and consultants at each of the stages.   KRI has proved its excellent technology management capabilities through its achievements in the past.   Both innovation and technology management are the key for success in R&D for new business management in the 21st century.   KRI, which incorporates these two functions, will, through its R&D activities, be expected to further produce major contributions.

12.   In Conclusion

While the influence of Japan's long recession has been quite large, during this time, the number of companies that can be counted as profitable is by no means small. Characteristics of these companies are originality in technology and products in addition to management style. Many of the innovative technologies were born in and grew under a rather unfavorable economy during which people felt uncomfortable or had a sense of something missing in their lives. Superb technological advances in computer software and biotechnology are overwhelming examples of some of these innovative technologies.

Now is the time for you to be a driving force in developing innovative technologies through positive R&D investment for your future. KRI sincerely hopes to continue serving as a group made up of quality professionals who embrace a dream, and to contribute to society through its achievements. At the same time, the company strives to be sensitive to changing needs, and, of course, to conduct R&D and consulting activities which will ensure success for its clients.

Macromol. Symp. **159**, 69–76 (2000)

Catalytic Diversity of Diene Complexes of Niobium and Tantalum on Polymerizations of Ethylene, Norbornene, and Methyl Methacrylate

Kazushi Mashima

Department of Chemistry, Graduate School of Engineering Science,
Osaka University, Toyonaka, Osaka 560-8531, Japan
E-mail: mashima@chem.es.osaka-u.ac.jp

SUMMARY: Half-metallocene diene complexes of niobium and tantalum catalyzed three-types of polymerization: (1) the living polymerization of ethylene by niobium and tantalum complexes, $MCl_2(\eta^4\text{-}1,3\text{-diene})(\eta^5\text{-}C_5R_5)$ (**1-4**; M = Nb, Ta; R = H, Me) combined with an excess of methylaluminoxane; (2) the stereoselective ring opening metathesis polymerization of norbornene by bis(benzyl) tantalum complexes, $Ta(CH_2Ph)_2(\eta^4\text{-}1,3\text{-butadiene})(\eta^5\text{-}C_5R_5)$ (**11**: R = Me; **12**: R = H) and $Ta(CH_2Ph)_2(\eta^4\text{-}o\text{-xylylene})(\eta^5\text{-}C_5Me_5)$ (**16**); and (3) the polymerization of methyl methacrylate by butadiene-diazabutadiene complexes of tantalum, $Ta(\eta^2\text{-}RN\text{=}CHCH\text{=}NR)(\eta^4\text{-}1,3\text{-butadiene})(\eta^5\text{-}C_5Me_5)$ (**25**: R = *p*-methoxyphenyl; **26**: R = cyclohexyl) in the presence of an aluminum compound (**24**) as an activator of the monomer.

Introduction

Olefin polymerization based on metallocene initiator system of Group 4 metallocene/methylaluminoxane (MAO) has been extensively investigated.[1,2] The polymerization using single-component catalysts such as cationic Group 4 metallocenes or isoelectronic neutral Group 3 and lanthanide metallocene hydrides or alkyls has recently been developed.[3-5] Furthermore, the polymerization based on non-metallocene catalysts has opened a new era.[6] The fragments of $MCp(\eta^4\text{-}1,3\text{-diene})$ (M = Nb and Ta) are isoelectronic and isolobal to those of MCp_2 (M = Zr and Hf).[7] This prompted us to prepare the benzyne complex, $Cp^*(\eta^4\text{-buta-}1,3\text{-diene})Ta(\eta^2\text{-}C_6H_4)$ (Cp* = η^5-pentamethylcyclopentadienyl),[8] and the benzylidene complex, $Cp^*(\eta^4\text{-buta-}1,3\text{-diene})Ta(\text{=}CHPh)(PMe_3)$,[9] whose structure and reactivity were compared with those of the corresponding metallocene complexes of Group 4 metals as well as those of metallocene-like complexes such as niobium complexes having $Nb(NR)Cp$ fragments.[10] We have investigated the diversity of the diene complexes of niobium and tantalum as catalyst precursors for the olefin polymerization; 14 electron isoelectronic species for Group 3, 4, and 5 metals being schematically shown in Scheme 1. This review briefly focuses on aspects of our past and current research on the polymerization.

Scheme 1: Active 14-electron species

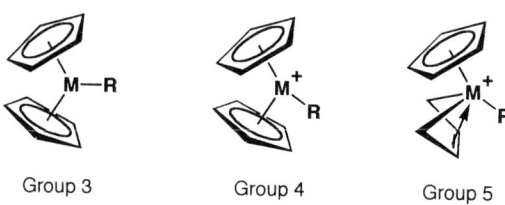

Group 3   Group 4   Group 5

Living Polymerization of Ethylene[11-13]

We used mono-diene complexes of niobium and tantalum, $MCl_2(\eta^5\text{-}C_5R_5)(\eta^4\text{-diene})$ (M = Nb and Ta; R = H and Me; diene = buta-1,3-diene, isoprene, 2,3-dimethylbuta-1,3-diene), as catalyst precursors of ethylene polymerization.   These complexes have been synthesized by the reaction of $MCl_4(\eta^5\text{-}C_5R_5)$ with the corresponding 1,3-diene compounds of magnesium.[14] Treatment of **2a-c** with 2 equiv. of MeMgI afforded dimethyl complexes of tantalum **5a-c**. In the case of the preparation of **6a** and **6c**, we used one-pot reaction of TaCl$_4$Cp with 2 equiv. of the corresponding methylated allyl Grignard reagent and 2 equiv. of MeMgI in THF, since the corresponding dichloro-diene complexes **4a** and **4c** have low solubility in organic solvent. A niobium dimethyl complex **7** was prepared similarly, but **7** was thermally unstable and decomposed gradually at room temperature to give a nascent carbene complex, which can be trapped by benzophenone to give 1,1-diphenylethylene.   Bis(2,3-dimethyl-1,3-butadiene) complexes **8** and **9** were prepared by the reaction of MCl_4Cp* with two equiv. of Mg(2,3-diemthyl-1,3-butadiene).[14]   All diene complexes mentioned above were the active catalyst precursors for ethylene polymerization when combined with 500 equiv. of MAO. Representative results using these mono-diene and bis(diene) complexes are shown in Table 1.

The dichloro and dimethyl complexes **2** and **5** exhibited almost the same catalyst activity, MAO thus inducing both the methylation and the ligand exchange reaction between aluminum and tantalum to lead the formation of cationic species.   The most attracting finding is the living nature of the polymerization; for example, using the niobium complexes **1a** and **1c** we obtained polyethylene with the narrowest polydispersity ($M_w/M_n = 1.05$) to date.   Other aspects of the catalysts are that the niobium complexes are superior to the corresponding tantalum complexes and that the bis(diene) complexes were also the active catalyst precursors for the ethylene polymerization.

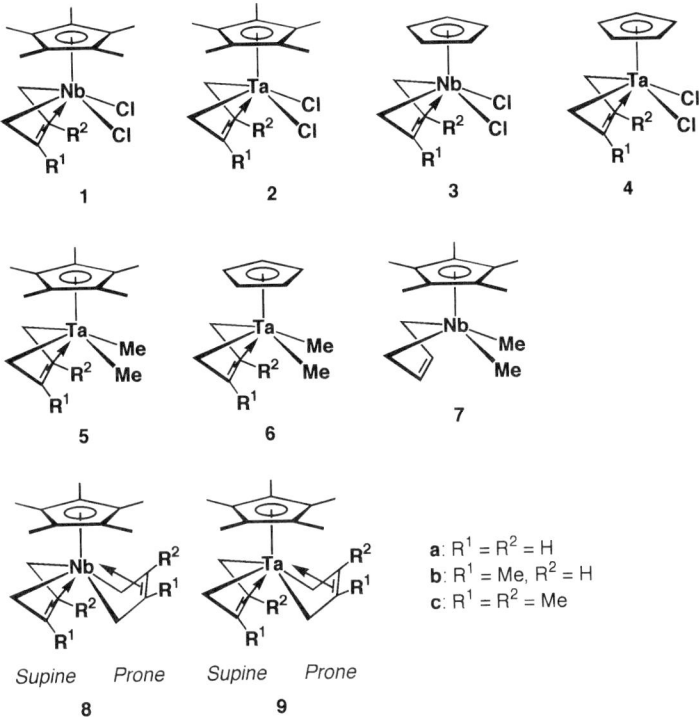

a: $R^1 = R^2 = H$
b: $R^1 = Me$, $R^2 = H$
c: $R^1 = R^2 = Me$

Table 1. Polymerization of Ethylene Catalyzed by Mono-diene Complexes/MAO.[a]

run	complex	time h	temp °C	activity[b] kg/h·[M]mol	$M_n/10^4$	M_w/M_n
1	1a	1	-20	10.65	2.36	1.05
2	1b	1	-20	1.02	0.51	1.09
3	1c	1	-20	12.71	4.10	1.05
4	2a	6	-20	1.51	2.03	1.16
5	4a	6	-20	7.07	8.18	1.40
6	5a	6	-20	1.18	2.55	1.08
7	8c	1	-20	12.48	1.42	1.06
8	9c	6	-20	0.26	0.54	1.09

[a]   Polymerization reactions were carried out in toluene ([M] = 1.44 x 10^{-3} M) in the presence of MAO (500 equiv.) for 6 hours without other notice.

[b]   For the methanol-insoluble parts.

In order to get an insight into the active species, we first examined the reaction of **5b** with 1 equiv. of TfOH in toluene, generating $TaCp^*(\eta^4\text{-isoprene})(OSO_2CF_3)Me$ (**10**) with elimination of 1 equiv. of methane. In the ^1H NMR spectrum of **10**, one set of signals was observed, indicating one of the two non-equivalent methyl groups selectively replaced by the TfO anion. As revealed by a crystallographic study, the complex **10** has the Ta-O covalent bond and thereby did not exhibit any activity in the polymerization. The addition of $B(C_6F_5)_3$ to **5b** generated a cationic species, despite its thermal instability, which catalyzed the polymerization.

Stereoselective ROMP of Norbornene[9,15,16]

We also studied the ring-opening metathesis polymerization (ROMP) of norbornene by using tantalum alkylidene species having $Ta(\eta^5\text{-}C_5R_5)(\eta^4\text{-butadiene})$ and $TaCp^*(\eta^4\text{-}o\text{-}$ xylylene) fragments,[9,15,16] because isoelectronic titanocene-alkylidene species have successfully been applied for the ROMP of norbornene.[17-19] Bis(benzyl) complexes **11** and **12** were prepared by the reactions of the dichloro complexes **2** and **4**, respectively, with 2 equiv. of benzyl Grignard reagent. The thermolysis of **11** and **12** in the presence of PMe_3 afforded the corresponding benzylidene phosphine complexes **13** and **14**, respectively. The phenyl group of the benzylidene moiety in the complexes **13** and **14** pointed away toward the $\eta^5\text{-}C_5R_5$ ligand; this *anti*-geometry may play a crucial role in determining the stereochemistry of the double bond of the poly(norbornene).

11: R = CH₃
12: R = H

13: R = CH₃
14: R = H

The *o*-xylylene ligand can coordinate to a metal center in η^4-fashion similar to the η^4-1,3-diene ligand. Thus, we anticipated that $Ta(CH_2Ph)_2(\eta^4\text{-}o\text{-}(CH_2)_2C_6H_4)Cp^*$ (**16**), which can be readily synthesized from **15**, initiates the ROMP of norbornene. The thermolysis of **16** produced a phosphine-free benzylidene complex, $Ta(=CHPh)(\eta^4\text{-}o\text{-}(CH_2)_2C_6H_4)Cp^*$ (**17**), whose crystallographic study revealed its *syn*-geometry around the Ta=C bond, being in sharp contrast to the *anti*-one found for **13** and **14**.

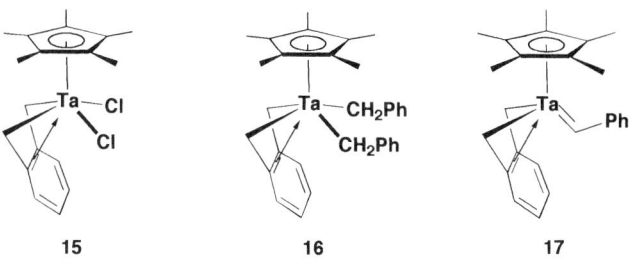

15   16   17

The *cis*-bis(benzyl) complexes **11**, **12**, and **16** together with the benzylidene complex **17** were found to be catalyst precursors for the ROMP of norbornene, whereas the isolated benzylidene-phosphine complexes **13** and **14** had no catalytic activity due to the coordination site occupied by the phosphine ligand. The stereochemistry of the vinylene double bonds of poly(norbornene) was effectively controlled by the catalyst precursor's architecture; **11** gave the polymer with a high *cis*-vinylene double bond (97-99%), while, in sharp contrast, **16** and **17** gave the polymer with a high *trans*-vinylene double bond (92-95%). This preferential difference is attributed to the geometry around the Ta=C bond of the nascent carbene species, *i.e.*, *anti*-rotamer or *syn*-one. The addition of norbornene to each rotamer resulted in the selective formation of metallacyclobutanes **18** and **20**. The metathesis cleavage of **18** affords the alkylidene species **19** along with the *cis*-double bond. The successive polymerization gives rise to the *cis*-poly(norbornene). The similar cleavage of the *o*-xylylene complex **20** leads to the *trans*-double bond and the *anti*-alkylidene species **21**, which turns the *syn*-one **22** before the successive addition of the monomer.

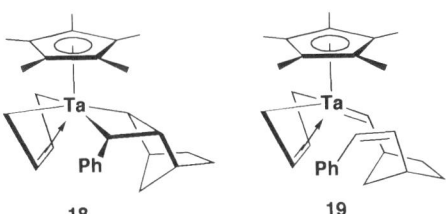

18   19

The evidence of metallacyclobutane intermediates was obtained by the crystallographic study for an acenaphthylene complex **23**, in which acenaphthylene added to the Ta=C bond in place of the norbornene and the stereochemistry of the *syn*-rotamer of **17** was retained (eq 1). The Cp complex **12** showed no stereoselectivity in double bond formation, indicating that the Cp and butadiene ligands exert similar steric effects and hence both *anti*- and *syn*-rotamers formed during the propagation.

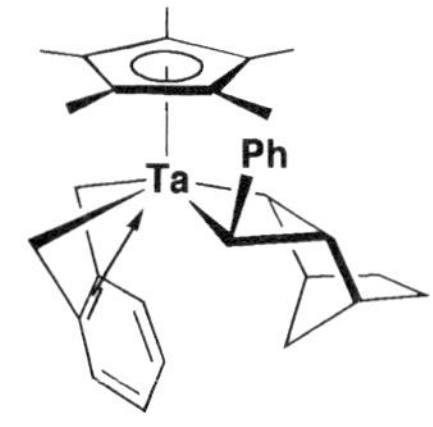

20 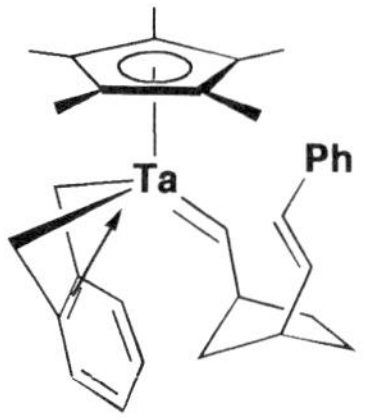 **21** 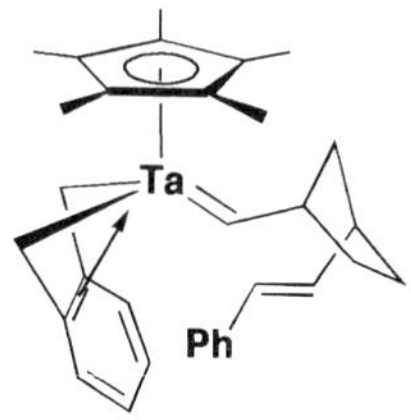 **22**

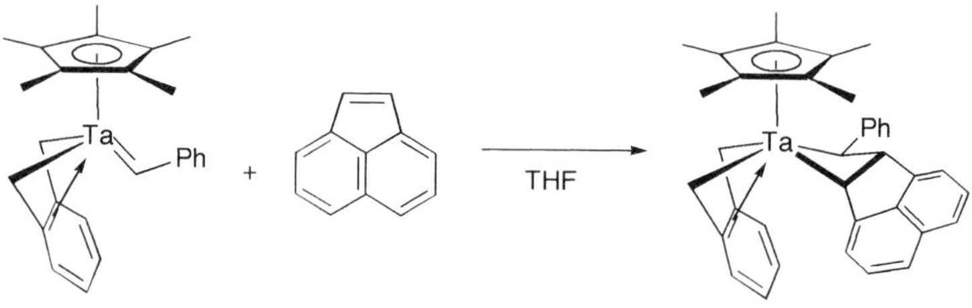

17 **23** (1)

Polymerization of Methyl Methacrylate[20]

The polymerization of functionalized olefins such as methyl methacrylate (MMA) using metallocene complexes of lanthanoid and zirconium has attracted recent interest.[21-28] On the basis of our above findings, it is assumed that half-metallocene-diene complexes of tantalum have the capability of polymerizing the polar monomers because the Group 5 metals are much tolerant to functionalized monomers.

In ethylene polymerization, the bis(diene) complexes of niobium and tantalum became unique catalyst precursors, in which one of the two diene ligands, *supine* and *prone* ones, was readily released upon treatment with MAO. Thus, the bis(diene) tantalum complex **9a** was first examined for the polymerization of MMA (Table 2), but it showed low activity. The polymerization of MMA proceeded much faster when combined with a bulky substituted bis(aryloxo)aluminum compound **24** as an activator of the monomer.[20]

1,4-Diaza-1,3-butadiene (= DAD) ligand is much tolerant to functionalized monomers and thereby the DAD-diene complexes of tantalum[29] were used as the catalyst precursors for the polymerization of MMA. Results summarized in Table 2 indicate that the DAD complexes **25** and **26** are superior in catalytic activity to the bis(diene) complex **9a**.[30] Further investigation to synthesize organometallic species active for the MMA polymerization is now in progress; various MMA complexes of tantalum were quite recently found to be excellent catalyst precursors for the polymerization of MMA.[30]

Ar = p-MeO-C_6H_4

24   **25**   **26**

Table 2. Polymerization of MMA Catalyzed by Diene Complexes of Tantalum.

run	catalyst	additive[a]	temp °C	time min.	yield[b] %	M_n[c] (x 10^4)	M_W/M_n[c]	rr
1	**9a**	-	20	1440	7	-	-	-
2	**9a**	**24**	0	40	98	29.3	2.3	68
3	**25**	**24**	0	40	99	44.2	2.5	73
4	**26**	**24**	0	10	81	14.4	1.5	71
5	**26**	**24**	0	5	99	16.9	1.4	70

a) [Cat.] : [MMA] = 1 : 100; [Cat.] : [**24**] : [MMA] = 1 : 10 : 100.

b) Yield = weight of polymer obtained/weight of monomer used.

c) Measured by GPC calibrated with standard polystyrene samples.

Conclusions

We demonstrated the catalytic diversity of the half-metallocene diene complexes of tantalum which catalyzed three different polymerizations (the living polymerization of ethylene, the stereoselective ROMP of norbornene, and the polymerization of methyl methacrylate) by controlling mutable auxiliary ligands, *i.e.*, chloro or methyl, diene, benzylidene, and 1,4-diaza-1,3-butadiene, bound to the tantalum center

Acknowledgment

I wish to thank my active co-workers and dedicated students listed in the reference section. This work was financially supported by the Grant-in-Aid for Scientific Research on Priority Areas (No. 283, "Innovative Synthetic Reactions") from the Ministry of Education, Science, Sports, and Culture, Japanese government, and also by the Yamada Science Foundation.

References

1   H.-H. Brintzinger, D. Fischer, R. Mühaupt, B. Rieger, R. M. Waymouth, *Angew. Chem., Int. Ed. Engl.* **34**, 1143 (1995).

2   W. Kaminsky, *J. Chem. Soc., Dalton Trans.* 1413 (1998).

3   K. Mashima, Y. Nakayama, A. Nakamura, *Adv. Polym. Sci.* **133**, 1 (1997).

4   M. Bochmann, *J. Chem. Soc., Dalton Trans.* 255 (1996).

5   R. H. Grubbs, G. W. Coates, *Acc. Chem. Res.* **29**, 85 (1996).

6   G. J. P. Britovsek, V. C. Gibson, D. F. Wass, *Angew. Chem., Int. Ed. Engl.* **38**, 428 (1999).

7   A. Nakamura, K. Mashima, *J. Organomet. Chem.* **500**, 261 (1995).

8   K. Mashima, Y. Tanaka, A. Nakamura, *Organometallics* **14**, 5642 (1995).

9   K. Mashima, Y. Tanaka, M. Kaidzu, A. Nakamura, *Organometallics* **15**, 2431 (1996).

10   V. C. Gibson, *J. Chem. Soc., Dalton Trans.* 1607 (1994).

11   K. Mashima, S. Fujikawa, A. Nakamura, *J. Am. Chem. Soc.* **115**, 10990 (1993).

12   K. Mashima, S. Fujikawa, H. Urata, E. Tanaka, A. Nakamura, *J. Chem. Soc., Chem. Commun.* 1623 (1994).

13   K. Mashima, S. Fujikawa, Y. Tanaka, H. Urata, T. Oshiki, E. Tanaka, A. Nakamura, *Organometallics* **14**, 2633 (1995).

14   H. Yasuda, K. Tatsumi, T. Okamoto, K. Mashima, K. Lee, A. Nakamura, Y. Kai, N. Kanehisa, N. Kasai, *J. Am. Chem. Soc.* **107**, 2410 (1985).

15   K. Mashima, M. Kaidzu, Y. Nakayama, A. Nakamura, *Organometallics* **16**, 1345 (1997).

16   K. Mashima, M. Kaidzu, Y. Tanaka, Y. Nakayama, A. Nakamura, J. G. Hamilton, J. J. Rooney, *Organometallics* **17**, 4183 (1998).

17   L. R. Gilliom, R. H. Grubbs, *J. Am. Chem. Soc.* **108**, 733 (1986).

18   L. R. Gilliom, R. H. Grubbs, *Organometallics*, **5**, 721 (1986).

19   N. A. Petasis, D.-K. Fu, *J. Am. Chem. Soc.* **115**, 7208 (1993).

20   K. Mashima, Y. Matsuo, K. Tani, *Proc. Japan Academy* **74**, 217 (1998).

21   H. Yasuda, E. Ihara, *Adv. Polym. Sci.*, **133**, 53 (1997).

22   H. Yasuda, H. Yamamoto, K. Yokota, S. Miyake, A. Nakamura, *J. Am. Chem. Soc.* **114**, 4908 (1992).

23   H. Yasuda, H. Yamamoto, M. Yamashita, K. Yokota, A. Nakamura, S. Miyake, Y. Kai, N. Kanehisa, *Macromolecules*, **26**, 7134 (1993).

24   M. A. Giardello, Y. Yamamoto, L. Brard, T. J. Marks, *J. Am. Chem. Soc.* **117**, 3276 (1995).

25   S. Collins, D. G. Ward, *J. Am. Chem. Soc.* **114**, 5460 (1992).

26   Y. Li, D. G. Ward, S. S. Reddy, S. Collins, *Macromolecules* **30**, 1875 (1997).

27   K. Soga, H. Deng, T. Yano, T. Shiono, *Macromolecules* **27**, 7938 (1994).

28   H. Deng, T. Shiono, K. Soga, *Macromolecules* **28**, 3067 (1995).

29   K. Mashima, Y. Matsuo, K. Tani, *Organometallics* **18**, 1471 (1999).

30   Unpublished results.

*Macromol. Symp. **159**, 77–88 (2000)*

Block Copolymers As Nanoscopic Templates

Thomas P. Russell*[a], Thomas Thurn-Albrecht[a], Mark Tuominen[b], Elbert

Huang[a], Craig J. Hawker[c]

[a] Department of Polymer Science and Engineering and [b] Department of
Physics and Astronomy, University of Massachusetts, Amherst, MA
01003, U S A
[c] IBM Almaden Research Center, 650 Harry Road, San Jose,
CA 95720; US A

SUMMARY: Block copolymers self-assemble into well-ordered,
microphase separated morphologies having dimensions on the molecular
scale. The key to the use of these nanoscopic structures lies in controlling
the spatial orientation of the morphology, particularly in thin films. The
preferential interactions of the segments of the blocks with interfaces
forces an alignment of the morphology parallel to the interface. Here we
describe the use of controlled interfacial interactions and electric fields to
manipulate the orientation of the morphology and subsequent steps
towards the generation of nanoporous templates as scaffolds for
nanoscopic structures.

Introduction

Polymeric materials offer a wealth of morphologies that span length scales ranging

from the submolecular to the macroscopic. In thin films of multicomponent polymer

systems, for example, block copolymers or polymer mixtures, making full use of

these morphologies requires precise control over the spatial orientation of the

morphology. This, in the absence of an external field, reduces to controlling the

interfacial energy between the phases, γ_{AB}, which is proportional to the segmental

interaction parameter, χ_{AB}; the interfacial energy between the components and the

substrates, γ_{AS} and γ_{BS}; the surface energies of the components, σ_A and σ_B; and the

commensurability between the film thickness, t, and the natural length scale of the

polymeric system[1-4]. For example, with an A-B diblock copolymer, the relative

magnitudes of σ_A and σ_B, will dictate which components segregate to the air surface.

The relative magnitudes of γ_{AS} and γ_{BS} will dictate which component resides

preferentially at the substrate interface. γ_{AB} or χ_{AB} will determine the energy required

     CCC 1022-1360/00/$ 17.50+.50/0

to sustain an interface between components A and B. Finally, the natural repeat period of the copolymer, i.e. the lattice constant, in comparison to the total film thickness will define whether the copolymer chains must stretch or compress or form a surface topography to accommodate an imposed frustration or whether the morphology will change its orientation with respect to the surface to minimize the free energy. If we examine each of these parameters for a specific A-B diblock copolymer, σ_A, σ_B, and χ_{AB} are dictated by the chemical composition of the copolymer and they are invariant. While the fundamental period of the copolymer is fixed, the film thickness of a specimen can readily be changed in the preparation. This leaves γ_{AS} and γ_{BS} as the only variables in the problem. By gaining a control over the interfacial interactions with the substrate, a means to manipulate the orientation of morphologies in thin films can be achieved.

Recently, a simple, robust route to control interfacial interactions was reported[5, 6]. Here, a random copolymer is anchored to a substrate by a terminal functional group. The synthesis of the random copolymer allows the composition of the copolymer to be varied from pure A to pure B or any fraction of A and B in the chain. The chemical variations along the chain are on the monomer or several monomers scale and, hence, the interaction experienced by an entire chain will reflect the average composition of the random copolymers, in particular, random copolymers of styrene and methylmethacrylate prepared by a "living" free radical polymerization with a terminal hydroxy group. These were anchored to a silicon oxide substrate. Contact angle measurements of thin polystyrene (PS) and poly(methylmethacrylate) (PMMA) films cast on the substrates modified with the random copolymer showed that there was one composition of the anchored random copolymer where the interactions of PS and PMMA were balanced. Block copolymers of PS and PMMA, denoted P(S-b-MMA), prepared on such "neutral" surfaces showed that the microdomain morphology of the copolymer oriented normal to the substrate or film plane. Thus, using this very simple surface modification, precise control over the orientation of the microdomains can be achieved.

In cases where such surface modification is not possible or where thicker films are required and the influence of the surface is lost due to defects, alternative methods must be used. In particular, an external field must be used to achieve the desired microdomain orientation. In the bulk, mechanical shearing has been shown to be the most effective means of orienting the microdomains of copolymers[7,8,9]. This, however, is not possible or practical in thin films. Any particulate impurities in a film preclude the possibility of maintaining two surfaces parallel over any reasonable lateral length scales. Electric fields, on the other hand, have been shown to be a viable means of orienting the microdomain structure in block copolymers[10,11]. However, in the bulk, it has been shown to be inefficient, with large voltages required to produce the desired orientation. More recently, other researchers[12,13] have shown that by placing the electrodes in close proximity, small applied voltages produce substantial fields and are quite effective in orienting the copolymer morphology. This work has recently been extended to thin films where the electrodes are placed on both surfaces of the film[14]. An additional contribution from the difference in the interfacial energies between the components enters into this problem and a threshold field strength is observed, above which, full alignment of the copolymer microdomain structure normal to the film surface is found.

Here, we report some recent results on the use of these controlled morphologies to produce novel, nanoporous media using block copolymers as templates. While there has been a tremendous effort expended in understanding the thermodynamics of the microdomain structure and there has been much research on copolymers justified by the potential applications of these structures for nanoscopic device application, little has been done to realize the potential. Here, it is shown that by coupling the control of the microdomain structure in thin films with standard methods to degrade one of the components, exceptional nanoscopic structures can be produced in a straightforward manner. Two examples will be discussed for P(S-b-MMA) using both controlled interfacial interactions and electric fields to achieve controlled microdomain orientation. The generation of nanoporous films will be shown and a brief discussion of possible uses of these structures will be presented.

Synthesis

P(S-r-MMA), with a styrene fraction of 0.60 (as determined by NMR), was synthesized in bulk via a TEMPO "living" free radical polymerization using a unimolecular initiator. This provided random copolymers with one hydroxy and one TEMPO terminus. The weight average molecular weight, M_W, and polydispersity, M_W/Mn, was determined to be 9,600 and 1.80, respectively, by size exclusion chromatography. The preparation of the unimolecular initiator and the polymerization of the random copolymer have been described previously[15-19].

Several asymmetric, diblock copolymers of polystyrene (PS) and poly(methylmethacrylate) P(S-b-MMA), having narrow molecular weight distributions with molecular weights ranging from $4-7 \times 10^4$ and with PMMA volume fraction of 0.29 were synthesized anionically. X-ray scattering was performed with Ni-filtered Cu-Kα radiation from a Rigaku rotating anode, operated at 8KW. A gas-filled area detector (Siemens Hi-Star) was used.

A layer of P(S-r-MMA) was anchored to a silicon substrate as described previously[5]. Excess random copolymer was washed away with toluene. The composition of the random copolymer was ~60% by volume of styrene, which, for PS and PMMA, represents a neutral surface. Onto this heated surface, a solution of

P(S-b-MMA) in toluene was spin coated. The total thickness of the film was ~40 nm. Without further surface treatment, the sample was heated to 170°C for 72 hrs under vacuum. Upon cooling to room temperature, atomic force microscopy studies clearly showed a hexagonally close-packed array of PMMA cylindrical microdomains at the surface.

Subsequent electron microscopy studies on fractured surfaces showed that the cylindrical microdomains extended from the air-surface to the substrate[6]. Consequently, the balancing of the interfacial interactions at the substrate interface, coupled with the film thickness, which places constraints on the microphase separated morphology[1-4], was sufficient to force the cylindrical microdomains to orient normal to the surface of the film.

This film was then exposed to UV light, which served two very important functions. First, PMMA is a well-known UV resist. Consequently, the PMMA will degrade upon exposure to UV light. Secondly, UV light crosslinks the PS matrix. This immobilizes the matrix, thereby, retaining the initial structure of the copolymer. The sample was washed with acetic acid and water to remove the degradation products of the PMMA. Two different field emission scanning electron micrographs are shown in Figure 1a and 1b. Figure 1a shows a view from the top of the sample, whereas Figure 1b shows a side view of the sample. There are several features of these images that are noteworthy. First, the morphology has been preserved. In fact, small angle x-ray scattering experiments have shown that there is absolutely no change in the separation distance between the cylindrical domains before and after removal of the PMMA. Secondly, the PMMA appears to have been completely removed from the sample with the cylindrical pores running from the surface to the substrate.

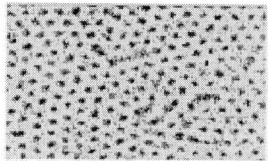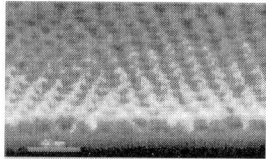

Fig. 1: (a) Field emission scanning electron micrograph (top view) of a nanoporous film made from an asymmetric P(S-b-MMA) diblock copolymer on a "neutral," random copolymer surface. (b) Field emission scanning electron micrograph (side view) of a nanoporous film made from an asymmetric P(S-b-MMA) diblock copolymer on a "neutral," random copolymer surface.

The films shown in Figures 1a and 1b offer numerous potentials for applications, either as scaffolds in which to build nanoscopic elements; as well-defined nanoporous media for separation purposes; or as templates to transfer the structure in the polymer film to the underlying substrate. It is this final application that will be demonstrated here.

The nanoporous film in Figure 1b was placed in a reactive ion etcher (RIE) using CF_4 as an etchant. The CF_4 will etch both the polymer template and the underlying silicon substrate. However, the presence of the polymer template on the surface will, of course, prevent the etching of the silicon underneath the polymer template initially. With time, the RIE etching will remove the polymer and begin etching the silicon. The exposed silicon substrate will already have been etched, however, thereby transferring the template into the substrate. A side-view FESEM image of the substrate is shown in Figure 2. The light region in the image is the silicon. What is evident from this image is that the pattern transfer into the silicon has clearly occurred. At the surface, one sees a very well-defined surface roughness with lateral feature sizes that are identical to the initial polymer film.

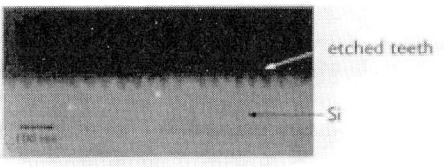

Fig. 2: FESEM image of a silicon substrate RIE etched with CF_4 where there was a copolymer template on the surface.

The image shown in Figure 2 is quite promising in terms of applications. It is, of course, clear that the aspect ratios must be increased to be of any commercial interest. Nonetheless, the effectiveness of the transfer is evident. What is more remarkable is the simplicity of the process. Within very few steps, a nano-textured surface in silicon can be generated in a very efficient and effective manner, relying only on the modification and neutralization of interfacial interactions.

The major limitation of this process is the ability to synthesize a random copolymer of the right composition to balance interfacial interactions. In some cases this may not be possible and, as such, an external field is required to achieve the desired orientation of the copolymer microdomains. Recently, we have shown that electric fields are a very effective means to this end. In these experiments, the diblock copolymer is sandwiched between two electrodes and the sample is annealed in the presence of a field. The anisotropic shape of the domains, coupled with the difference in the dielectric constants of the components comprising the domains should orient the mirodomains in the direction of the field lines.

Shown in Figure 3 are three transmission electron micrographs of a P(S-b-MMA) copolymer having PMMA cylindrical microdomains annealed between two electrodes. The substrate (one electrode) was a gold layer that was evaporated onto an Ultem poly(etheretherimide) plaque. This substrate facilitates subsequent microtoming of the sample. The second electrode was an aluminized Kapton sheet that was simply placed on the surface of the copolymer.

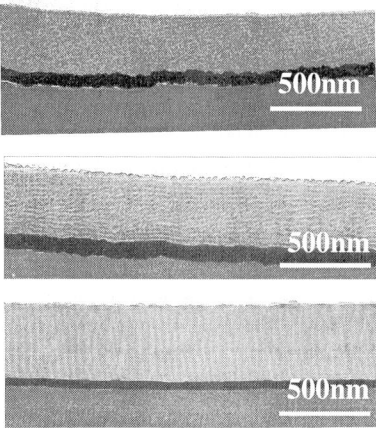

Fig. 3: (a) TEM micrograph of an asymmetric P(S-b-MMA) diblock copolymer having cylindrical microdomains. The film was cast on a gold substrate (dark band in film) and annealed with no electric field. (b) Same as in 3a, however, a field strength of several V/µm was applied during annealing. (c) Same as in 3a, however, a field strength of several 10V/µm was applied during annealing.

The uppermost image in Figure 3a is that of a microtomed P(S-b-MMA) film after annealing without any applied external field. The microphase separated structure of the copolymer is evident though the alignment of the microdomains parallel to the substrate is poor, due to the sample thickness and the slight preference of PS to segregate to the substrate. Upon the application of a low field, Figure 3b, a rather startling result was observed. Rather than aligning parallel to the surface, the cylindrical microdomains aligned normal to the field lines, i.e. parallel to the substrate surface. This result was precisely opposite to the expected result and suggests that a partial polarization of the microdomains occurs and the interaction between the microdomains enhances the ordering. At higher field strengths, Figure 3c, the cylindrical microdomains are seen to align parallel to the field lines, i.e.

normal to the surface. The cylindrical microdomains are seen to extend from the substrate through the sample to the air interface. More detailed studies have shown that orientation of the microdomains does not occur until a threshold field strength is exceeded[14]. This threshold field strength is greater than the field strength required to orient the diblock copolymer in the bulk. The difference between the field strengths required to align the bulk and thin film field strength is directly related to the difference in the interfacial energies of the two components with the substrate.

A very convenient means of probing the orientation of the cylindrical microdomains in the polymer films in a nondestructive manner is by use of x-ray scattering. To make sure of x-ray scattering, the substrate must be transparent or semi-transparent to the x-rays. To this end P(S-b-MMA) films were cast onto a 75μm thick Kapton film coated with a thin layer of aluminum (0.1μm in thickness). A similar Kapton film served as the upper electrode. After annealing the film at 180°C under a field of 24V/mm, the x-ray beam was passed through the sample at an incidence angle of 45°. Weak reflections characteristic of the oriented copolymer morphology were observed. At small values of the scattering vector $q(=(4\pi/\lambda)\sin(\theta)$ where λ is the wavelength and 2θ is the scattering angle), the scattering in the equatorial direction decreased monotonically with a small maximum characteristic of the oriented PMMA cylinders. Since the film is only 1.3μm in thickness, the total scattering volume is small. In addition, the electron density difference between PS and PMMA, is not very large. Both factors contribute to the weak yet clearly observable scattering maximum. Exposing the copolymer film to UV radiation, washing with acetic acid and drying produced a dramatic change in the scattering. The two dimensional scattering pattern shown in the l.h.s. of Figure 4 shows two intense reflections along the equator. The dramatic increase in the scattering is clearly seen in the azimuthal scans of the intensity at a fixed-q corresponding to the maximum. Shown in the r.h.s. of Figure 4 is a comparison of the azimuthal scans before and after irradiation. The ratio in the peak intensity before and after the UV treatment shows an increase in the scattering by a factor of 55. If the UV treatment has generated a porous structure then the scattering would increase by a factor of $(\rho_{PMMA}-\rho_{PS}-\rho_{PMMA})^2$ where ρ_i is the electron density of component i. This calculated factor is 53, in quantitative agreement with the observed increase in the scattering. These results show that the PMMA cylinders have been quantitatively removed from the film while maintaining the structure characteristic of the initial block copolymer film. If we consider the film thickness of $1.3\mu m$ then the pores have an aspect ratio of over 100! AFM measurements along with other electrochemical experiments clearly show that these cylinders extend from the substrate interface all the way to the air surface.

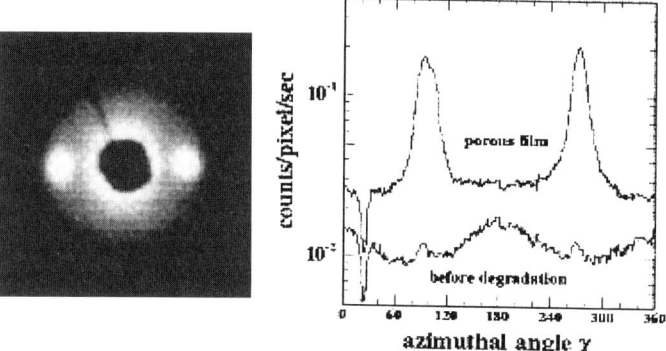

Fig. 4: (l.h.s). Two-dimensional SAXS pattern of a porous film made from an oriented P(S-b-MMA) copolymer having cylindrical PMMA microdomains. (r.h.s.) Comparison of azimuthal scans intensity for a P(S-b-MMA) diblock copolymer before and after degradation of the PMMA microdomains.

Thus, the two examples shown demonstrate easy, robust methods for generating nanoporous structures in thin films with well-defined orientations of the cylinders with respect to the surface. While one can easily imagine that such structures are ideal candidates for separations media, it is worthwhile to consider some of the other possible areas that such structures may impact. With the nanoporous films on a conducting surface, the film can be placed in an electrochemical bath and metal can easily be deposited within the scaffold defined by the porous film. This indeed has been done with cobalt, a magnetic material, forming an ordered array of magnetic nanowires in the film. Alternatively, catalysts can be anchored to the exposed substrate and polymerization can be conducted within this confined geometry. Functional polymer chains can, also, be anchored to the exposed substrate which forms the basis of a sensing array.

It is evident that the possible uses of such structures are numerous. However, the key elements to enabling such applications and potential commercialization are that one can manipulate the self-assembling structures of polymers in thin films in an easy, robust and reproducible manner. It is precisely this which has been described here.

Acknowledgments

We are grateful to C. Black and L. Raboin for their help with the FESEM. This work was supported by the U.S. Department of Energy, Office Basic Energy Sciences under contract of DE-FG02-96ER45612, the National Science Foundation Partnership in Nanotechnology under CTR-9871782, the NSF Materials Research Science and Engineering Center (DMR-9809365), and the National Institute of Standards and Technology, U.S. Department of Commerce, in providing the neutron research facilities. T. T.-A. acknowledges the support of the Deutsche Forchungsgemeinschaft.

References

1. D. G. Walton, G. J. Kellogg, A. M. Mayes, P. Lambooy, T. P. Russell, *Macromolecules,* **27**, 6225-6228 (1994)
2. M. S. Turner, *Phys. Rev. Lett.,* **69**, 1788 (1992)

3. M. Kikuchi, K. Binder, *J. Chem. Phys.,* **101**, 3367-3377 (1994)
4. M. W. Matsen, *J. Chem. Phys.,* **106** (1997)
5. P. Mansky, Y. Liu, E. Huang, T. P. Russell, C. Hawker, *Science*(1458) (1997)
6. E. Huang, T. P. Russell, C. Harrison, P. M. Chaikin, R. A. Register, C. J. Hawker, J. Mays, *Macromolecules,* **31**, 7641-7650 (1998)
7. I. W. Hamley (1998). The Physics of Block Copolymers. Oxford, Oxford University Press
8. Z.-R. Chen, J. A. Kornfield, S. D. Smith, J. T. Grothaus, M. M. Satkowski, *Science,* **277**, 1248 (1997)
9. D. L. Polis, K. I. Winey, *Macromolecules,* **29**, 8180 (1996)
10. K. Amundson, E. Helfand, D. D. Davis, X. Quan, S. S. Patel, *Macromolecules,* **24**, 6549 (1991)
11. K. Amundson, E. Helfand, X. Quan, S. D. Hudson, S. D. Smith, *Macromolecules,* **27**, 6559 (1994)
12. T. Morkved, M. Lu, A. M. Urbas, E. E. Ehrichs, H. M. Jaeger, P. Mansky, T. P. Russell, *Science,* **273**, 931 (1996)
13. P. Mansky, J. DeRouchey, T. P. Russell, J. Mays, M. Pitsikalis, T. Morkved, H. M. Jaeger, *Macromolecules,* **31**, 4399 (1998)
14. T. Thurn-Albrecht, T. P. Russell, H. M. Jaeger, *Macromolecules* (submitted)
15. M. K. Georges, R. P. N. Veregin, P. M. Kazmaier, G. K. Hamer, *Macromolecules,* **26**, 2987 (1993)
16. I. Li, B. A. Howell, K. Matyjaszewski, T. Shigemoto, P. B. Smith, D. B. Priddy, *Macromolecules,* **28**, 6692 (1995)
17. C. J. Hawker, *J. Am. Chem. Soc.,* **116**, 11314 (1994)
18. C. J. Hawker, E. Elce, J. Dao, W. Volksen, T. P. Russell, G. G. Barclay, *Macromolecules,* **29**, 2686 (1996)
19. C. J. Hawker, G. G. Barclay, A. Orellana, J. Dao, W. Devonport, *Macromolecules,* **29**, 5245 (1996)

Surface Aggregation Structure and Surface Mechanical Properties of Binary Polymer Blend Thin Films

Atsushi Takahara[*], Kensuke Nakamura[**], Keiji Tanaka[**] and Tisato Kajiyama[**]

[*]Institute for Fundamental Research of Organic Chemistry and [**]Graduate School of Engineering, Kyushu University, Higashi-ku, Fukuoka 812-8581, JAPAN

SUMMARY: During preparation of very thin polymer belnd films from a solution of polymers, the phase-separated structures which are quite different from that observed for the bulk blend film was observed. From atomic force microscopic(AFM) observation, it is concluded that the surface undulation, which reflects the phase separated morphology of the blend system, is present. In the case of (polystyrene(PS)/poly(methyl methacrylate)(PMMA)) blend system, a large influence of end-group chemistry on the surface morphology was observed. The phase identification of the (rubbery polymer/glassy polymer) binary blend thin films was successfully achieved by scanning vioscoelasticity microsopy(SVM).

Introduction

Surface structure and properties of multiphase polymers have been paid a great attention these days because surface structure and properties are closely related to many technological applications[1,2]. As predicted by the mean-field theory, the polymer blend thick films showed the enrichment of lower surface free energy component at the air/solid interface[1,2]. In the case of polymer blend thin films, the molecular aggregation state can be expected to be different from that of the thick film due to the strong influence of the air and substrate interfaces. In this study, the surface structure and surface mechanical properties of various binary polymer blend systems have been studied on the basis of scanning force microscopy(SFM).

Experimental

Polymer blend thin films were prepared in order to image the surface structure and surface mechanical properties at the polymeric solid. Polymer blend systems used in this study were (polystyrene(PS)/poly(vinyl methyether) (PVME)), (PS/polyisoprene(PI)) and (PS/poly(methyl methacrylate)(PMMA)). The bulk (PS/PVME) blend is miscible state at room temperature, whereas the (PS/PI) and (PS/PMMA) blend systems are in an immiscible state at room temperature. Table 1 summarizes the physico-chemical properties of polymers used in this study for blends and the preparation methods. Unless specified the

 CCC 1022-1360/00/$ 17.50+.50/0

Table 1   Characterization of blend components

Blend	Blend Component	Mn	Mn/Mw	Tg/K	γ/mN m^{-1}	Preparation Method
(PS/PVME)	PS	26.6k	1.09	386	40.2	Dip-coating
	PVME	28.8k	1.98	249	36.0	
(PS/PI)	PS	55k	1.09	386	40.2	Spin-coating
	PI	52k	1.18	200	36.0	
(PS/PMMA)	PS	54.5	1.09	386	40.2	Spin-coating
	PMMA	46.3k	1.12	399	44.0	
(α,ω-PS(NH$_2$)$_2$/PMMA)						
	α,ω-PS(NH$_2$)$_2$	53.4	1.08	386	40.2	Spin-coating
	PMMA	46.3k	1.12	399	44.0	

chain end groups of PS and PMMA are composed of *sec*-butyl group and proton-terminated styrene unit, respectively. The diamino-terminated PS[α,ω-PS(NH$_2$)$_2$] was synthesized by a living anionic polymerization. The blend ratio of (Polymer A/Polymer B) was designated as (weight% /weight%). In order to avoid an influence of the surface segregation of low molecular weight component, the blends were prepared from monodisperse polymers except for (PS/PVME) blend.

AFM was used for   the observation of the surface morphology of the blend films.   The AFM images were obtained by an SPA 300 with SPI 3700 controller (Seiko Instruments Industry Co., Ltd.) at 293 K in air.   The cantilever used for AFM observation was microfabricated from Si$_3$N$_4$ and its spring constant was 0.02 Nm^{-1}.   The two-dimensional mapping of dynamic mechanical properties for the phase-separated surface of binary polymer blends were carried out by utilizing scanning viscoelasticity microscope (SVM)[3).   When the cantilever tip is positioned in a repulsive-force region of the force curve, the sample surface might be deformed by the indentation of the tip.   The modulation of the indented tip leads to the modulation of the force between sample surface and tip.   If the modulation is applied sinusoidally, the dynamic viscoelastic properties at the sample surface can be evaluated by measuring the amplitude of the modulated deformation of the sample and the phase lag, between modulation signal and modulated deformation of the sample.   The SVM measurement and observation were performed at 293 K in air under a repulsive force region. The modulation frequency and the modulation amplitude were 4 or 5 kHz and 1.0 nm, respectively.   The cantilever used was microfabricated from Si$_3$N$_4$ and its spring constant was 0.09 Nm^{-1}.

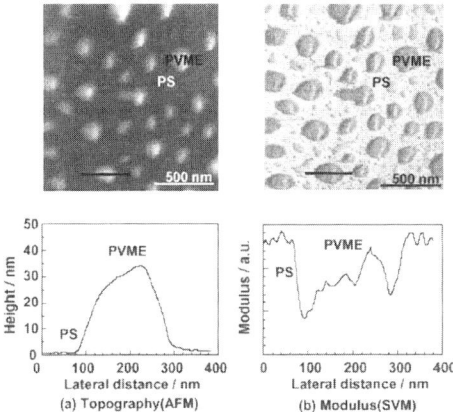

Fig.1: AFM and SVM images of (PS/PVME) (62/38 w/w) blend thin films with thickness of ca.25nm.   Height profiles along a line are also shown.

Results and Discussion

(PS/PVME) Blend Ultrathin Films

[Polystyrene/poly(vinyl methyl ether)](PS/PVME) is a blend system with a lower critical solution temperature (LCST) type phase diagram.   At room temperature, this blend system shows a phase mixing state in the bulk state.   X-ray photoelectron spectroscopy (XPS) of the blend thick film showed the enrichment of lower surface free energy PVME at the surface. However, the surface enrichment of PVME became less prominent with a decrease in film thickness[4].   Infrared spectroscopic measurement revealed that the (PS/PVME) blend ultrathin film of ca.25 nm thick prepared from a (50/50 w/w) toluene solution by a dip-coating method has the blend ratio of (PS/PVME) (62/38 w/w) in a bulk region due to a selective adsorption of PS segments to the hydrophilic substrate. Fig.1 shows AFM and SVM images of (PS/PVME) (62/38) blend thin films with thickness of ca.25nm. Even if the observation temperature was below the bulk cloud point, AFM observation revealed that the (PS/PVME) ultrathin film with thickness of ca.25nm was in an apparent phase-separated state in which the droplet-like domains of 200-500 nm in diameter and 20-40 nm in height were formed. The characterization of the droplet-like domains was carried out using AFM observation of the (PS/PVME) ultrathin film with different (PS/PVME) blend ratios.   The apparent surface area of the droplet-like domains on the AFM image decreased with a decrease in the PVME weight fraction.   Then, it is apparent from the AFM observation that the droplet-like domains are composed of the PVME rich phase. Scanning viscoelasticity microscopic (SVM) observation was carried out in order to evaluate surface viscoelasticity of the as-cast (PS/PVME) (62/38) ultrathin film of ca.25 nm thick on the hydrophilic substrate[4].   Since the bulk T_g of PS is far above room temperature, whereas, that of PVME is below room temperature, it is expected that the glassy PS and the rubbery PVME phases can be distinguished apparently even at the surface on the basis of the SVM observation. Since droplet-like domain showed lower modulus than the matrix phase, it seems reasonable to conclude that the droplet-like domains are composed of the rubbery PVME rich phase and that, the matrix is composed of the glassy PS one. The contrast in the AFM image reflects the difference of the sample height, whereas, that in the SVM image comes from the difference of the modulus. Thus, the combination of AFM and SVM images can reveal the interfacial characteristics of the two-phase system even though the interfacial region between domain and matrix is not distinct.

(PS/PI) Blend Thin Films

The surface of the (PS/PI) film, which is a typical phase-separated blend system, was investigated by using AFM and SVM. In the case of the bulk blend film, a macroscopic phase separation was observed. Fig.2 shows the AFM and SVM images of the as-cast (PS/PI) (70/30 w/w) blend thin film of 100 nm thick prepared by spin-coating[5]. AFM image showed the well-defined phase-separated structure with the domains of ca. 2 μm in diameter and the apparent domain height was ca. 70 nm lower than matrix. On the other hand, in the case of the blend ratio of (30/70), it was observed in the AFM image that the domain was ca. 70 nm higher than matrix. Since the apparent area of the brighter part in the AFM image increased with the blend ratio of PS, it seems reasonable to conclude that the higher height region at the phase-separated surface is composed of the PS rich phase which has higher surface free energy component in this blend system. Also, when a surface etching treatment of the (PS/PI) (30/70) blend film was carried out using methylethylketone, which is a good solvent only for PS, the domains were disappeared from the surface. Therefore, this result indicates again that the domains and the matrix regions of (PS/PI)(30/70w/w) blend thin film are composed of the PS and PI, respectively. The formation of high PS-rich domain is closely related to the large surface tension of PS compared with that of PI. SVM observation was carried out at room temperature, which was far below and above Tgs of PS and PI, respectively. Since the lower modulus region in the SVM image corresponds to the lower height regions in AFM image, it is reasonable to conclude that the higher height region is composed of the glassy PS rich phase and also, the lower height region are composed of the rubbery PI one. The domain radius in the AFM image was almost the same as that observed

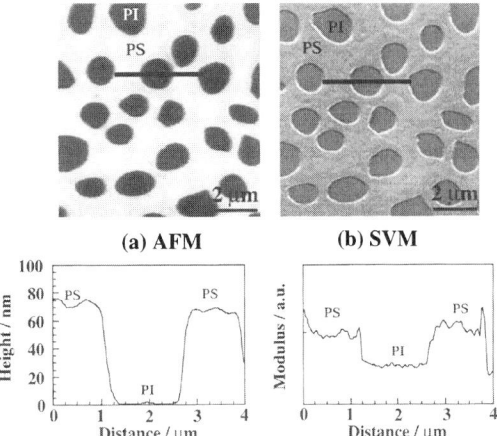

(a) AFM   **(b) SVM**

Fig.2: AFM and SVM images of (PS/I) (50/50) blend thin films with thickness of 100nm. Height profiles along a line are also shown.

by SVM.   This would mean the fairly distinct interfacial boundary, in other words, a small interfacial thickness between the PS and PI phases.   Since the (PS/PI) blend system has larger magnitude of interaction parameter, χ_{12} compared with that of (PS/PVME) system, the interfacial thickness must be much narrower than that of (PS/PVME) system.

(PS/PMMA) Blend Thin Films

It has been reported that the chain end groups which have a lower surface free energy than that of the main chain unit are enriched at the surface and that the thermal molecular motion at the surface is activated due to an increase in free volume fraction at the surface[6].   On the other hand,   the chain end groups,   which have a higher surface free energy than that of main chain unit, are depleted from the surface to minimize the interfacial free energy[7]. However, little study has been done on the control of phase separation based on the interaction between polymer chain end groups and substrate.   The monodisperse proton-terminated PS(Mn=54.5k), diamino-terminated PS[α,ω-PS(NH$_2$)$_2$] (Mn=53.4k) and proton-terminated PMMA(Mn=46.3k) were used to prepare the polymer blend thin films. The (PS/PMMA) blend thin films were prepared onto a silicon wafer with native oxide layer by the spin-coating method from their toluene solution at several spin speeds.   The change in spin speed corresponds to the change in the rate of solvent quench. The blend films were vacuum-dried for 24 h at room temperature. An apparent equilibrium aggregation structure at the blend thin film surface was obtained by annealing thin films at 423 K for 24 h and/or 48 h.

Phase-separated structure of the (PS/PMMA) blend thick film, whose thickness is 25 μm, was investigated to compare with that of the thin film.   The AFM and the phase contrast

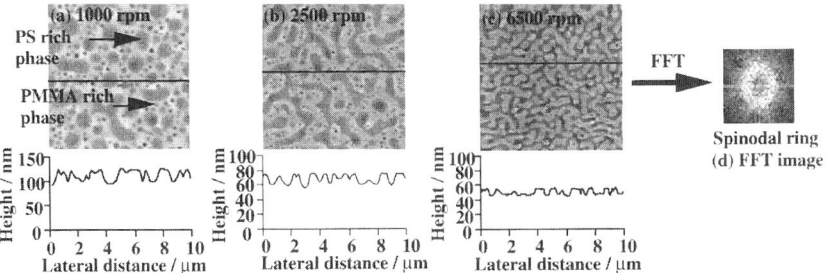

Fig.3: (a)-(c) AFM images and line profiles of the (PS/PMMA) (20/80) thin spin-coated at (a) 1000, (b) 2500 and (c) 6500 rpm and (d) FFT image.

microscopic (PCM) images of the (PS/PMMA) film of 25 μm thick revealed that although the surface was not in an apparent phase-separated state, a macroscopic phase-separated structure was formed in a bulk. XPS measurement revealed that the PS was enriched at the film surface due to its lower surface free energy compared with that of PMMA. In order to reveal the influence of film thickness on surface morphology, the surface structures of the (PS/PMMA) thin films with various blend ratios were observed with AFM. Figure 3(a) shows the AFM image of the (PS/PMMA) (20/80) thin films of ca.100 nm thickness spin-coated at 1000 rpm. It was confirmed by surface etching treatment with cyclohexane that the higher regions were the PMMA-rich phase in all the (PS/PMMA) thin films. Fig.3 (b)-(c) shows the AFM images for the (PS/PMMA) (20/80) blend thin films of ca.100nm thickness with several spin speeds. The thickness of the film decreased with an increase in spin speed. The bicontinuous-like phase-separated structures were observed in all the (PS/PMMA) (20/80) blend thin films. The size of the phase-separation decreased with an increase in spin rate. Fig.3(d) shows the fast Fourier transform (FFT) image for the (PS/PMMA) (20/80) thin film prepared at 6500 rpm. A spinodal ring due to the periodic structure was observed in the FFT image. The phase-separated structure for the (PS/PMMA) (10/90) and (70/30) blend thin films were the reverse sea-island-like and the sea-island-like phase-separated structures, respectively. The bright area in AFM image is smaller than that expected from the bulk composition. This behavior can be attributed to an increase in the apparent surface PS fraction due to a selective adsorption of PMMA to the silicon substrate. The AFM image of (PS/PMMA) blend system after extensive annealing revealed that the phase-separated structure is not stable because of the negative spreading coefficient of PS against PMMA.

The phase-separated structure for the $(\alpha,\omega\text{-PS(NH}_2)_2/\text{PMMA})$ thin films were investigated in order to reveal the influence of the chain end group chemistry on the surface morphology. AFM observation revealed that the bicontinuous-like phase-separated structure was observed

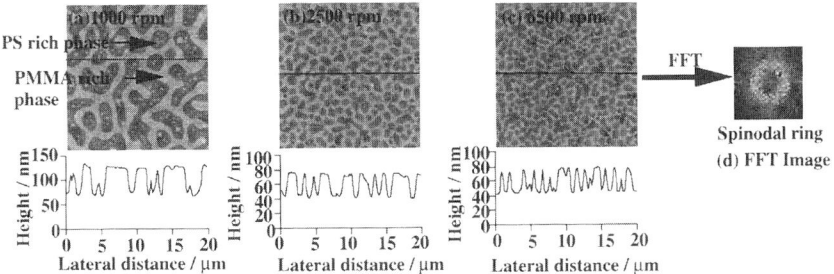

Fig.4: (a)-(c) AFM images and line profiles of the for (α,ω-PS(NH$_2$)$_2$/PMMA) (50/50)thin spin-coated at (a) 1000, (b) 2500 and (c) 6500 rpm and (d) FFT image.

at a PS fraction of 50wt%. The inverse sea-island-like and the sea-island-like phase-separated structures were formed in the case of the (α,ω-PS(NH$_2$)$_2$/PMMA) (30/70) and (70/30) thin films, respectively. Both AFM and XPS revealed that the surface PS composition increased with an increase in bulk PS composition. These results indicate that no preferential adsorption of the both components onto the hydrophilic silicon substrate occurred.

Fig. 4 (b)-(c) show the AFM images for the (α,ω-PS(NH$_2$)$_2$/PMMA) (50/50) thin films spin-coated at several spin speeds. The bicontinuous-like phase-separated structures were observed in all AFM images for the (α,ω-PS(NH$_2$)$_2$/PMMA) (50/50) thin films. The size of the phase separated domains decreased with an increase in spin rate. Fig. 4(d) shows the FFT image for the (α,ω-PS(NH$_2$)$_2$/PMMA) (50/50) thin film at 6500 rpm spin speed. A spinodal ring was observed in the FFT image. The presence of periodic structure suggests that the blend thin film was formed via spinodal decomposition during the solvent evaporation process. The phase-separated structure of the (α,ω-PS(NH$_2$)$_2$/PMMA) thin film was stable even after annealing above Tg for 25 hours. These results indicated that the phase separation of polymer blend thin film was greatly influenced by the preferential adsorption of one component to the substrate because the area ratio of air/polymer and polymer/substrate interfaces to the total volume of thin film is greater than that of the thick film.

The higher height regions for the (PS/PMMA) and (PS/PI) blend prepared by spin-coating methods are always higher surface energy component. Figure 5 schematically shows the formation mechanism of the height difference of binary immiscible polymer blend systems during the solvent evaporation process[8]. Polymer B with lower surface free energy tends to cover the air-polymer interfacial region in order to minimize the interfacial free energy. However, since the time required for the film formation is fairly short due to a very fast evaporation of the solvent and the limitation of the diffusion of the confined molecular chain,

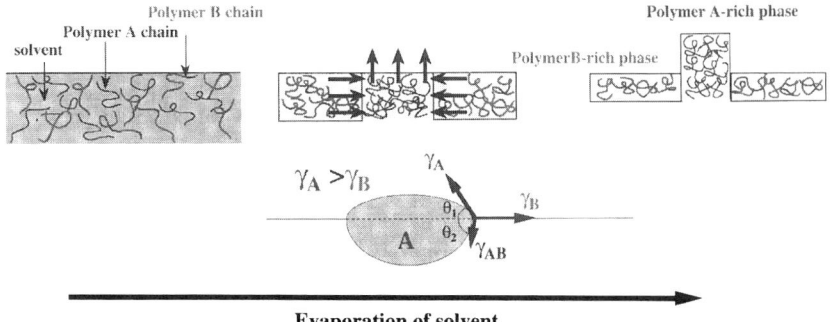

Evaporation of solvent

Fig.5: Schematic representation of the formation mechanism of the height difference of binary immiscible polymer blend systems.

the surface containing both polymer A and polymer B components with a little residual solvent was formed.   After film was formed, a local rearrangement of the surface occurs due to the presence of the residual solvent.   The higher height polymer A phase was formed owing to the large surface tension acting at the direction to the polymer A phase from the triphase boundary and the spreading of lower surface free energy polymer B component in order to minimize the interfacial free energy.

Conclusion

The phase-separated structures which are quite different from that for the bulk blend film was observed for the polymer blend thin films prepared from polymer solution. Atomic force microscopic(AFM) observation revealed the surface undulation which reflects the phase separated morphology of the blend system. In the case of (PS/PMMA) blend system, a large influence of end-group chemistry on the surface morphology was observed. The phase identification of the (rubbery/glassy) binary polymer blend thin films was successfully achieved by scanning vioscoelasticity microscopy(SVM).

References

1. R. A. L. Jones and R. W. Richards, *Polymer at Surface and Interfaces*, Cambridge University Press, Cambridge (1999)
2. A. Takahara, Chapt 7 in *Modern Approaches to Wettability Theory and Applications*, M. Schrader and G. Loeb Eds., Plenum, NY (1992)
3. T. Kajiyama, K. Tanaka, I. Ohki, S-R. Ge, J.-S. Yoon, A. Takahara, *Macromolecules*, **27**, 7932(1994)
4. K. Tanaka, J.-S. Yoon, A. Takahara, and T. Kajiyama, *Macromolecules*, **28**, 4, 934 (1995)
5. A. Takahara, X.-Q. Jiang, N. Satomi, K. Tanaka, T. Kajiyama, *Interfacial Aspects of Multicomponent Polymer Materials*, Plenum,NY 63(1997)
6. K. Tanaka, A. Taura, S.-R. Ge, A. Takahara, and T. Kajiyama, *Macromolecules*, **29**, 8, 3040 (1996)
7. K. Tanaka, X.-Q. Jiang, K. Nakamura, A. Takahara, T. Kajiyama, T. Ishizone, A. Hirao, and S. Nakahama, *Macromolecules*, **31**, 5148 (1998)
8. K. Tanaka, A. Takahara, and T. Kajiyama, *Macromolecules*, **29**, 3232 (1996)

Interfaces in polymer blends

Marcus Müller and Kurt Binder

Institut für Physik, WA 331, Johannes Gutenberg Universität
D-55099 Mainz, Germany
(IPC99, Yokohama, Macromolecular Symposia)

Summary: We investigate the structure and thermodynamics of interfaces in dense polymer blends using Monte Carlo (MC) simulations and self-consistent field (SCF) calculations. For structurally symmetric blends we find quantitative agreement between the MC simulations and the SCF calculations for excess quantities of the interface (e.g., interfacial tension or enrichment of copolymers at the interface). However, a quantitative comparison between profiles across the interface in the MC simulations and the SCF calculations has to take due account of capillary waves. While the profiles in the SCF calculations correspond to intrinsic profiles of a perfectly flat interface the local interfacial position fluctuates in the MC simulations. We test this concept by extensive Monte Carlo simulations and study the cross-over between "intrinsic" fluctuations which build up the local profile and capillary waves on long (lateral) length scales.

Properties of structurally asymmetric blends are exemplified by investigating polymers of different stiffness. At high incompatibilities the interfacial width is not much larger than the persistence length of the stiffer component. In this limit we find deviations from the predictions of the Gaussian chain model: while the Gaussian chain model yields an increase of the interfacial width upon increasing the persistence length, no such increase is found in the MC simulations. Using a partial enumeration technique, however, we can account for the details of the chain architecture on all length scales in the SCF calculations and achieve good agreement with the MC simulations.

In blends containing diblock copolymers we investigate the enrichment of copolymers at the interface and the concomitant reduction of the interfacial tension. At weak segregation the addition of copolymers leads to compatibilization. At high incompatibilities, the homopolymer-rich phase can accommodate only a small fraction of copolymer before the copolymer forms a lamellar phase. The analysis of interfacial fluctuations yields an estimate for the bending rigidity of the interface. The latter quantity is important for the formation of a polymeric microemulsion at intermediate segregation and the consequences for the phase diagram are discussed.

Introduction

Melt blending of polymers is a promising route for tailoring materials to specific application properties. Unlike metallic alloys, however, chemically different polymers often do not mix on microscopic length scales. Rather a complicated morphology of droplets of one component dispersed into the other component forms on a mesoscopic length scale, and the blend can be conceived as an assembly of interfaces. While the detailed structure on this mesoscopic length scale depends strongly on the way the material is processed, the local properties of interfaces are certainly crucial for understanding the material properties. The interfacial width sets the length scale on which entanglement between polymers of the different components form. Experiments[1] suggest that the mechanical strength in-

© WILEY-VCH Verlag GmbH, D-69469 Weinheim, 2000   CCC 1022-1360/00/$ 17.50+.50/0

creases if the interfacial width exceeds the entanglement length. The interfacial tension is important for the breaking–up of droplets under shear[2,3]: The lower the interfacial tension the finer the dispersed the two components.

In the following we shall describe computer simulations in the framework of a coarse grained lattice model targeted at investigating the local structure of the interfaces between coexisting phases in polymer blends. The results are compared to self-consistent field calculations and a description via an effective interface Hamiltonian. In the next section we describe our computational model and the Monte Carlo technique. Then we detail our results on symmetric binary blends, binary blends of polymers with different stiffness, and a ternary blend of two homopolymers and a symmetric diblock copolymer. The paper closes with a short outlook.

Model and technique

The simulations are performed in the framework of the bond fluctuation model. In the framework of this coarse grained lattice model in three dimensions each monomer represents a small number of repeat units (say, 3-5) along the backbone of a polymer. Only the relevant interactions – excluded volume, connectivity, and a repulsion between unlike species – are retained. Each monomer blocks 8 corners of a unit cell of the simple cubic lattice from further occupancy. We work at a number density of $\rho = 1/16$. At this density the model reproduces many features of a dense melt, e.g., the screening of the excluded volume, but still allows for a fast relaxation of the chain conformations. Monomers are connected via one of 108 bond vectors of length $2, \sqrt{5}, \sqrt{6}, 3$, and $\sqrt{10}$ in units of the lattice spacing. The large number of bond vectors permits 87 distinct bond angles and yields a good approximation of continuous space. The bending stiffness of a polymer can be tuned by imposing a bending potential $E(\Theta) = f \cos(\Theta)$, where Θ denotes the angle between successive bond vectors. It acts only on the B component. The blend comprises two species – A and B – which interact via a square well potential of depth ϵ. The contact of like monomer in the range of the interaction $r \leq \sqrt{6}$ lowers the energy by an amount ϵ whereas the contact of unlike monomers increases the energy by ϵ. The energy scale of the mutual interactions sets the temperature scale.

In our model the thermal interactions can be related to the Flory–Huggins parameter via the intermolecular pair correlation function $g^{\text{inter}}(r)$. This quantity measures the probability of finding a monomer of another chain at a distance r. Clearly, only the interactions between different chains contribute to the energy of mixing, and we obtain:

$$\chi = \frac{2\epsilon}{k_B T} \int_{r \leq \sqrt{6}} \mathrm{d}\mathbf{r}^3 \; g^{\text{inter}}(\mathbf{r}) \approx 5.3\epsilon \tag{1}$$

where the last expression holds for flexible chains. Entropic contributions are small compared to the enthalpic ones for the parameters considered in the following. A more detailed account of the thermodynamics of the homogenous system can be found in Ref[4].

Unless noted otherwise we use chain length $N = 32$ in our simulations. This corresponds roughly to a degree of polymerization of 100-150. Two types of simulation schemes have been employed to investigate interfacial properties: simulations of the interface in the canonical ensemble and simulations in the semi-grandcanonical ensemble in junction with reweighting techniques. In the semi–grandcanonical ensemble the composition of the mixture fluctuates and the chemical potential difference $\Delta\mu$ between the polymer species is held fixed. The Monte Carlo Scheme comprises two types of movements: Canonical moves which relax the conformations of the polymers on the lattice and switches of the chain identity $A \rightleftharpoons B$. In the latter scheme the Hamiltonian, which is simulated, is modified as to encourage the system to sample configurations which contain interfaces. In this way all excess quantities of the interface (e.g., interfacial tension, interfacial energy, or the enrichment of copolymers) are accurately measurable in the simulations. The canonical simulations can also take advantage of the fast relaxation of identity switches by exchanging the identity between pairs of polymers.

The Monte Carlo simulations are compared to self-consistent field (SCF) calculations. Either the Gaussian chain model is used (and the chain dimensions are identified via the end–to–end distance R_e as measured in the MC simulations), or the chain conformations are extracted from the simulation of a bulk system and the single chain partition function, required by the SCF scheme, is approximated by a large (more than 7×10^7) sample of single chain conformations. The latter scheme has advantages if the interfacial width becomes comparable to the persistence length of the polymer.

Results

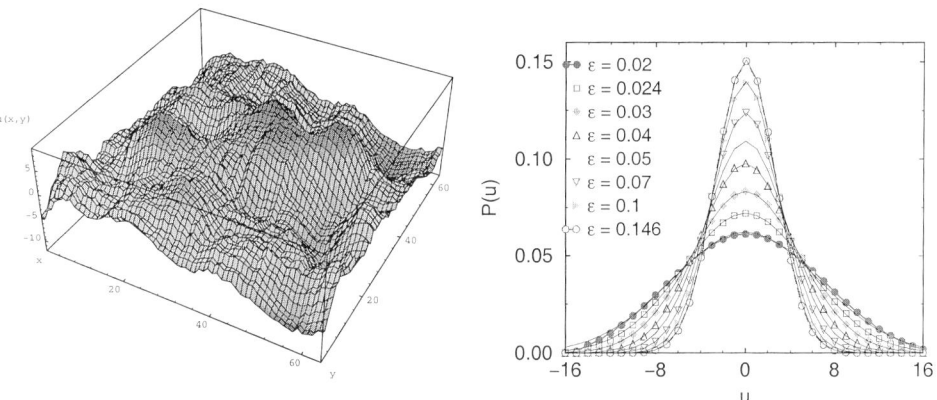

Figure 1: (a) Fluctuations of the local interface position in a binary polymer blend at $\epsilon = 0.03$ ($\chi N \approx 5.1$). The position has been averaged over a lateral size of $B = 8$. (b) Distribution of the local interface positions as a function of the incompatibility ϵ. From Werner et al.[5].

A comparison between simulations/experiments and SCF calculations has to take due account of the fluctuations of the local interface position. Interfaces are not perfectly flat, as assumed in the SCF calculations, but there are thermal fluctuations. A snapshot of the interface position in the MC simulations of a binary blend is shown in Fig.1(a). To a first approximation the effect of these fluctuations is to increase the effective area of the interface. Let $u(\mathbf{r}\|)$ denote the local interface position. Then, the free energy cost of deviations from perfectly flat configurations are described by the Hamiltonian $\mathcal{H}$:

$$\mathcal{H}[u(\mathbf{r}_\parallel)] = \int d^2\mathbf{r}_\parallel \left\{ \frac{\sigma}{2}|\nabla u|^2 + \frac{\kappa}{2}|\triangle u|^2 \right\} \tag{2}$$

This is an expansion in terms of small u and its derivatives, σ denotes the interfacial tension and κ the bending rigidity. For interfaces between not too long homopolymers κ is very small; for copolymer loaden interfaces, however, the second term becomes important as we shall discuss. This capillary wave Hamiltonian is diagonal and quadratic in terms of the Fourier components $u(q)$ and the equipartition theorem yields for the spectrum of fluctuations in thermal equilibrium:

$$\langle u^2(q) \rangle = \frac{k_B T}{\sigma q^2 + \kappa q^4} \tag{3}$$

The local positions $u(\mathbf{r}_\parallel)$ are also Gaussian distributed $P(u)$ with variance s:

$$s^2 = \frac{1}{4\pi^2} \int d^2\mathbf{q}_\parallel \langle u^2(\mathbf{q}_\parallel) \rangle = \frac{k_B T}{2\pi\sigma} \ln\left(\frac{q_{\max}}{q_{\min}}\right) \tag{4}$$

where a short and long length scale cut–off $q_{\max}$ and $q_{\min}$ have to be introduced to avoid the divergence at $q \to 0$ and $q \to \infty$. The bending rigidity κ has been neglected, it would make the cut–off at small distance obsolete. The MC result for the distribution $P(u)$ of the local positions is presented in Fig.1(**b**). Upon decreasing the incompatibility ϵ, we increase the strength of the fluctuations.

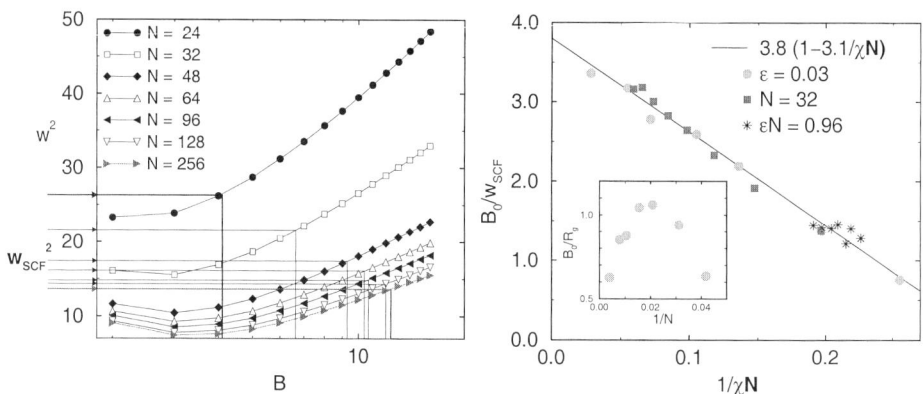

Figure 2: (**a**) Dependence of the apparent interfacial width on the lateral block size B at constant incompatibility $\epsilon = 0.03$ for different chain lengths. From the intersections of the simulation data with the self-consistent field predictions (presented as horizontal lines) we determine the small length scale cut–off B_0. (**b**) The ratio B_0/w_{scf} approaches a constant value 3.8. Leading corrections are of the order $1/\chi N$. Note that data sets at constant N, χN and χ collapse in this representation. The inset shows the ratio B_0/R_e for fixed incompatibility $\epsilon = 0.03$. The ratio tends to zero for large N. From Werner et al.[6].

These capillary waves broaden the apparent profiles p_{app}. Laterally averaged profiles, as obtained in experiments or MC simulations, are describable via the convolution of an intrinsic profiles p_{int} of an ideally flat interfaces and the distribution of the local positions

$$p_{\text{app}}(z) = \int du\, P(u) p_{\text{int}}(z - u) , \tag{5}$$

where z denotes the coordinate perpendicular. When applied to a erfc-shape profile, one obtains[5,7]:

$$w_{\mathrm{app}}^2 = w_{\mathrm{int}}^2 + \frac{k_B T}{4\sigma} \ln\left(\frac{q_{\max}}{q_{\min}}\right) \tag{6}$$

The apparent width is broader than the intrinsic one and depends via the two cut-offs on the system geometry[5]. For a free interface the upper cut–off is set by the lateral block size B on which the interface is observed. This might be set by the size of the simulation cell or the coherence length of the neutron beam by which the interfacial structure is investigated. Gravitation or interactions with walls/surfaces also limit long wavelength fluctuations. On short distances the separation of fluctuation into "intrinsic" ones, which build up the smooth profile of the ideally flat interface, and capillary waves breaks down. Polymer blends are well suited to examine this crossover, because the strong interdigitation of the molecules makes self-consistent field calculations describe the properties of interfaces accurately except for capillary wave fluctuations. Taking the SCF prediction as the width of an hypothetical flat interface, we have used Eq.6 to determine the length scale $B_0 = 2\pi/q_{\max}$ on which the crossover between the intrinsic fluctuations and capillary waves occurs. This procedure is illustrated in Fig.2(a). There are three possible candidates for B_0: A microscopic length scale (e.g., the bond length) independent from temperature or chain length; the width of the interface, which depends on temperature but not on N; or the radius of gyration/correlation length which depends both on χN and R_e. The simulation data[6] in panel (b) indicate a behavior of the form $B_0 = 3.8 w_{\mathrm{scf}}(1 - 3.1/\chi N)$, i.e. the intrinsic width of the interface sets the crossover length; a result compatible with calculations of Semenov[8].

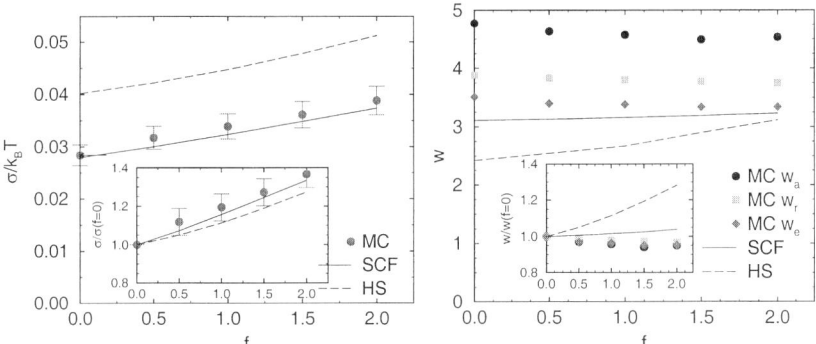

Figure 3: Interfacial tension (a) and interfacial width (b) in a blend of flexible ($f = 0$) and stiff (f as indicated) polymers at rather strong segregation $\epsilon = 0.05$. Comparisons with detailed SCF calculations, which take account of the chain architecture on all length scales, and predictions of the Gaussian chain model by Helfand and Sapse (HS) are shown. In panel (b) w_a denotes the apparent width, which is averaged over the whole lateral system size $L = 64$, w_r represents the width on the block size $B = 16$, and $w_e = 2\Delta e/\chi k_B T \rho$ denotes the width extracted from the excess energy Δe of the interface per unit area, respectively. From Müller and Werner[6].

The spectrum of interfacial fluctuations is a route for measuring the interfacial tension in MC simulations. This is illustrated for a blend with a structural asymmetry in Fig.3. When we increase the stiffness disparity, the semi-grandcanonical identity switches become increasingly inefficient and σ cannot be obtained via reweighting techniques. Upon increasing the stiffness, we increase the χ parameter (c.f. Eq.1), because the stiffer polymer has a more open structure and interacts with more neighboring monomers on different chains. Stiffness disparity gives also rise to an entropic contribution to χ, due to differences in the packing structure of the monomers[4]. For the rather large thermal interactions $\epsilon = 0.05$ this effect is, however, negligible. The increased incompatibility results in a larger interfacial tension (c.f. Fig.3(a)). This effect is quantitatively captured by the SCF calculations which account for the detailed chain architecture. Qualitative agreement is also obtained within the Gaussian chain model. The deviations from the prediction of Helfand and Sapse[10] are mainly due to chain end corrections. The situation is qualitatively different for the intrinsic width of the interface (c.f. panel (b)). Monte Carlo simulations and SCF calculations predict no increase or even a reduction of the width for larger stiffness, while the intrinsic width increases in the Gaussian chain model. The break down of the Gaussian chain model can be qualitatively rationalized as follows: The width of the interface is determined by loops of the polymers into the other phase. For large incompatibility the width of the interface becomes comparable with the persistence length and the conformation of a loop differs from the Gaussian statistic of the chain on large length scales. Likewise, MC simulations, SCF calculations and predictions of the Gaussian chain model agree for smaller χ, where $w \gg b$.

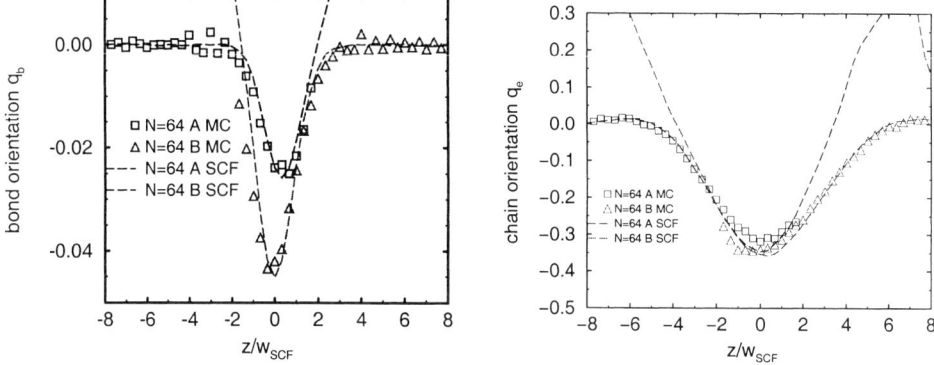

Figure 4: Orientations of the bond vector (**a**) and $\vec{R}_e$(**b**) in a blend of polymers with stiffness $f = 0$ (left side) and $f = 1$ (right side) at rather strong segregation $\epsilon = 0.05$ and chain length $N = 64$. Symbols denote the results of the MC simulation on a lateral length scale $B = 16$, while lines represent the SCF calculations. From Müller and Werner[9].

The ability of the SCF calculations to provide a detailed description of the intrinsic interface structure is illustrated in Fig.4. We present the MC and SCF results for the orientations as measured by the 2nd Legendre polynom of the angle between the bond vector or end-to-end distance $\vec{R}_e$ with respect to the interface. Both vectors align parallel to the interface, but the effect is much stronger for $\vec{R}_e$ and more stiffness dependent for the bond vectors.

Diblock copolymers are model surfactants for the AB homopolymer blend. They adsorb at the interface as to extend both halves into the corresponding homopolymer phases. This

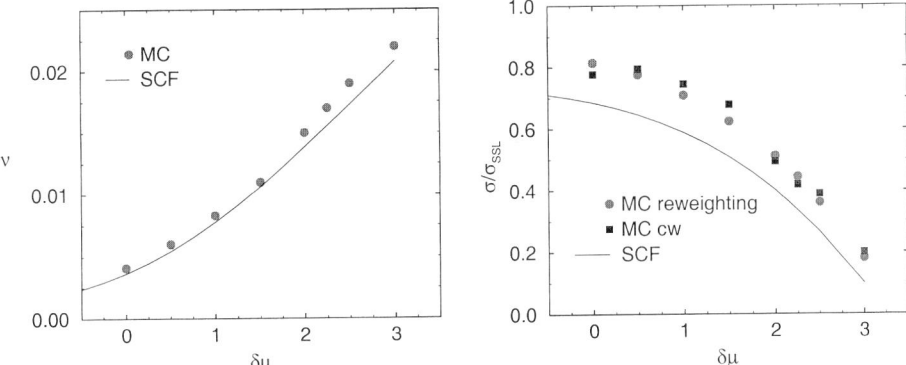

Figure 5: (a) Adsorption of diblock copolymers at a homopolymer/homopolymer interface as a function of the chemical potential of the copolymer at $\epsilon = 0.1$ and chain length $N = 32$. (b) Reduction of the interfacial tension upon adding copolymers. From Werner et al.[11].

decreases their enthalpy, but the localization at the interface reduces the translational entropy and the conformational entropy due to chain stretching at high copolymer excess at the interface. Upon increasing the chemical potential $\delta\mu$ (or concentration) of the copolymers in the bulk, we observe the adsorption of copolymers at the interface and the concomitant reduction of the interfacial tension in Fig.5. Both MC simulations and SCF calculations agree at high segregation. However, rather than forming a dense copolymer brush at the interface, a phase separation into a homopolymer–rich phase and a lamellar phase (swollen by homopolymers) is encountered. For small ϵ the addition of copolymers

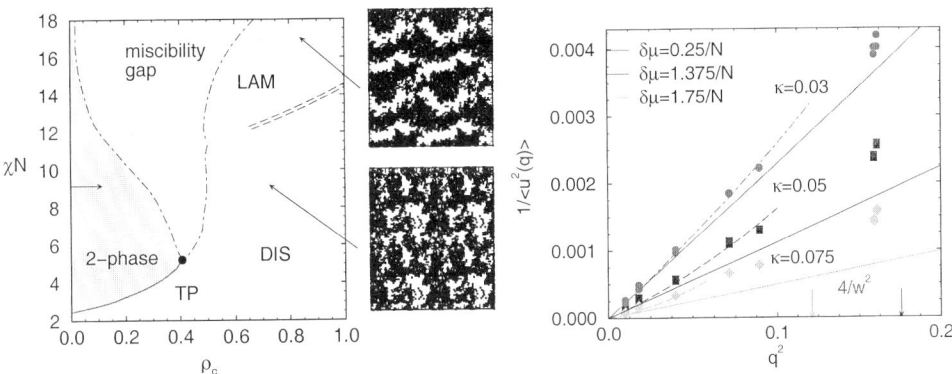

Figure 6: (a) Isopleth cut through the phase triangle of a ternary blend. (b) Spectrum of interface fluctuations in a ternary blend of two homopolymers and a diblock copolymer at $\epsilon = 0.054$. Estimates for the bending rigidity of the interfaces are indicated. From Müller and Schick[12].

drives the system to compatibility (c.f. Fig.6(a)). At intermediate segregation we find a three phase coexistence between two homopolymer–rich phases and a copolymer–rich disordered phase. The latter has a structure of a microemulsion (as revealed, e.g., by the snapshots). SCF calculations by Janert and Schick[13] rather predict highly swollen lamellar phases in this region. Some insight into this discrepancy can be gained from the spectrum of interface fluctuations. Upon adding copolymers to the systems at $\epsilon = 0.054$ we decrease the interfacial tension and deviations from a simple q^2 dependence become apparent (see panel (**b**)). We can obtain a rough estimate of the bending rigidity κ of the copolymer-loaden interface according to Eq.3. The bending rigidity turns out to be much smaller than $k_B T / 2\pi$. It is this bending rigidity, however, which stabilizes the liquid–crystalline order of the lamellar phase. De Gennes and Taupin[15] argued that a small value of κ leads to the formation of a microemulsion. Indeed this is observed in the simulation[12] and experiments[14]. If we were to increase the chain length we would increase the bending rigidity $\kappa \sim \sqrt{N}$[16] and stabilize the lamellar phases predicted by the SCF theory.

Outlook

Interfaces in polymer blends have been investigated by simulations and the results have been compared to SCF calculations. Special attention has been focused on interface fluctuations. They broaden interfacial profiles, which are measured in experiments or simulations, and they lead to the formation of polymeric microemulsions. They are also important for the wetting behavior and the properties of confined systems[17].

Acknowledgment: It is a pleasure to thank F. Schmid, A. Werner, and M. Schick for fruitful and enjoyable collaboration. Financial support from the DFG under grant Bi314/17 and access to the CRAY T3Es at the NIC Jülich, the HLR Stuttgart and the San Diego SC are acknowledged.

1. C. Creton, E.J. Kramer, and G. Hadziioannou, Macromolecules **24**, 1846 (1991)
2. G.I. Taylor, Proc.R.Soc. London, **A 138**, 41 (1932)
3. S.T. Milner, MRS Bull. **22**, 38 (1997)
4. M. Müller, Macromol.Theory Simul. **8**, 343 (1999)
5. A. Werner, F. Schmid, M. Müller, and K. Binder, J.Chem.Phys. **107**, 8175 (1997)
6. A. Werner, F. Schmid, M. Müller, and K. Binder, Phys.Rev. **E 59**, 728 (1999)
7. M.D. Lacasse, G.S. Grest, and A.J. Levine, Phys.Rev.Lett. **80**, 309 (1998)
8. A.N. Semenov, Macromolecules **27**, 2732 (1994)
9. M. Müller and A. Werner, J.Chem.Phys. **107**, 10764 (1997)
10. E. Helfand and A.M. Sapse, J.Chem.Phys. **62**, 1329 (1975)
11. A. Werner, F. Schmid, and M. Müller, J.Chem.Phys. **110**, 5370 (1999)
12. M. Müller and M. Schick, J.Chem.Phys. **105**, 885 (1996)
13. P.K. Janert and M. Schick, Macromolecules **30**, 3916 (1997)
14. G.H. Fredrickson and F.S. Bates, Polym.Sci. **B 35**, 2775 (1997)
    H.S. Jeon, J.H. Lee, and N.P. Balsara, Phys.Rev.Lett. **79**, 3274 (1997)
15. P.G. de Gennes and C. Taupin J.Phys.Chem. **86**, 2294 (1982)
16. M.W. Matsen, J.Chem.Phys. **110**, 4658 (1999)
17. M. Müller and K. Binder, Macromolecules **31**, 8323 (1998)

Macromol. Symp. **159**, 105–112 (2000)

Thermodynamic Characterization of Polymer-Polymer Interfaces

G. D. Merfeld and D. R. Paul*

Department of Chemical Engineering and Texas Materials Institute

The University of Texas at Austin, Austin, TX 78712, U S A

SUMMARY: The properties of multiphase polymer blends are determined in part by the nature of the polymer-polymer interface. The interfacial tension, γ, influences morphology development during melt mixing while interfacial thickness, λ, is related to the adhesion between the phases in the solid blend. A quantitative relation between the thermodynamic interaction energy and these interfacial properties was first proposed in the theory of Helfand and Tagami and has since been correlated with experimental measurements with varying degrees of success. This paper demonstrates that the theory and experiment can be unified for polymer pairs of some technological importance: copolymers of styrene and acrylonitrile (SAN) with poly (2, 6-dimethyl-1, 4-phenylene oxide) (PPO) and with bisphenol-A polycarbonate (PC). For each pair, the overall interaction energy was calculated using a mean-field binary interaction model expressed in terms of the interactions between repeat unit pairs extracted from blend phase behavior. Predictions of γ and λ as a function of copolymer composition made by combining the binary interaction model with the Helfand-Tagami theory compare favorably with experimental measurements.

Introduction

The structure and properties of a blend of polymers A and B are determined in large measure by the energetic interaction between A and B. When the interaction is more favorable than a critical value, the mixture forms a single homogeneous phase; however, when the interaction is less favorable than this critical value, the mixture separates into A-rich and B-rich phases. The boundaries (in terms of temperature, molecular weight, copolymer composition, etc.) between miscibility and immiscibility, or the phase diagram, are governed by the thermodynamic energy of interaction. In addition, the nature of the interface between the A-rich and B-rich phases in immiscible mixtures is also governed by this thermodynamic interaction energy. Thus, if suitable theories exist, experimental determination of appropriate phase diagram information should allow predictions of interfacial properties. This interconnection would be useful since measurements of interfacial tension and thickness are generally more difficult than establishing the boundaries between miscibility and immiscibility. The following

© WILEY-VCH Verlag GmbH, D-69469 Weinheim, 2000

summary of recent observations demonstrates that this interrelationship can be made quite successfully, for at least two systems of some importance, using existing theories [1-3].

Theoretical Framework for Establishing Polymer-Polymer Interactions

The phase behavior of mixtures is governed by the Gibbs free energy of mixing and how it depends on concentration, temperature, and the nature of the components. The simplest such expression for mixing a unit volume of monodisperse polymers A and B is the classical result

$$\Delta g_{mix} = B\phi_A\phi_B + RT\left[\frac{\rho_A\phi_A\ln\phi_A}{M_A} + \frac{\rho_B\phi_B\ln\phi_B}{M_B}\right] \tag{1}$$

where B is a binary interaction energy density, R is the gas constant, T is the absolute temperature, and ρ_i, ϕ_i, and M_i are the density, volume fraction and molecular weight of component i, respectively. This equation combines a Hildebrand-Scatchard-van Laar type heat of mixing with an expression for the combinatorial entropy of mixing given by Flory and Huggins [4,5]. In equation 1, B is an excess free energy term in which the heat of mixing plus other noncombinatorial effects are lumped. A parameter χ can be used in place of B and is related by the expression

$$\chi = \frac{BV_{ref}}{RT} \tag{2}$$

where V_{ref} is a reference volume which usually is taken as the molar volume of one of the repeat units in the system. However, χ is not well suited for current purposes because its value is dependent on the arbitrarily (and sometimes ambiguously) defined V_{ref}. Therefore, the use of the B parameter, which Flory identified as sometimes preferred when dealing with heterogeneous polymers [6], is strongly recommended. The relative balance between the energetic and entropic contributions to mixing determine whether the mixture forms one or two phases. The critical value of B is given by

$$B_{critical} = \frac{RT}{2}\left(\sqrt{\frac{\rho_A}{(\overline{M}_W)_A}} + \sqrt{\frac{\rho_B}{(\overline{M}_W)_B}}\right)^2 \tag{3}$$

where $\left(\overline{M}_W\right)_i$ is the weight average molecular weight of component i [7,8]. Miscibility exists when B is less than this critical value while phase separation occurs when B is larger than this value.

A simple mean field approximation for the quantitative representation of polymer-polymer interactions [9-11] has proven to be extremely useful for blends of copolymers when the interactions are not highly specific. This binary interaction model resolves interactions down to the repeat unit or structural segment level and considers pair-wise interactions that may occur between different polymers (*inter*molecular) and within a copolymer (*intra*molecular). The model assumes that polymer interactions can be represented in terms of the interactions that exist between repeat units and that the interactions of a given segment are unaffected by the neighboring segments to which it is covalently joined. Also, since the model draws upon the statistics of a mean field approximation to determine the probability of pair-wise interactions, it naturally assumes random mixing. This requires the copolymers themselves to be relatively random with respect to the distribution of segment types along their chain structures. Strictly speaking, the use of a mean-field approximation excludes specific interactions such as hydrogen bonding that may create spatial or directional ordering which would compromise the random mixing approximation; however, many such systems have been modeled with this simple formalism. The most general form of this model for mixing two polymers A and B containing an unlimited number of repeat units is given by

$$B = \sum_{i>j}^{\text{Inter}} B_{ij}\phi_i\phi_j - \sum_{i>j}^{\text{Intra}} B_{ij}\phi_i\phi_j \tag{4}$$

This reduces to the following special case

$$B = B_{13}\phi_1 + B_{23}\phi_2 - B_{12}\phi_1\phi_2 \tag{5}$$

for mixing a copolymer of units 1 and 2 with a homopolymer comprised of 3 units where ϕ_i refers to the volume fraction of i units in the polymer.

Polymer blend phase behavior can be predicted or analyzed by inserting the binary interaction model into the thermodynamic framework of either the Flory-Huggins theory or an appropriate equation-of-state theory. The simplicity of the Flory-Huggins theory makes it useful for evaluation of isothermal phase boundaries, miscibility maps, or UCST type phase separation. To treat LCST type phase separation without resorting to empiricisms of the Flory-Huggins theory, it is necessary to employ an equation-of-state analysis. In using either of these theories, it is simplifying to assume, for experimental and computational ease, that phase observations are made at the spinodal condition. This assumption is assured for blends at or near the critical composition where the binodal and spinodal curves overlap. Typically, the blend critical composition is about 50/50 by weight for polymers of roughly similar molecular weight. But for blends whose components differ significantly in molecular weight, the critical composition shifts toward the lower molecular weight component; the adjusted critical composition can be calculated as shown elsewhere [8].

Although polymer blend phase behavior is determined predominately by the nature of the molecular interactions, it may also be influenced by many other factors including molecular weight, copolymer composition, blend ratio, temperature and pressure. Systematically changing one of these variables or a combination of them and studying the effect produced on phase behavior affords an opportunity to characterize the relevant binary interaction parameters. Of the variables mentioned, molecular weight, copolymer composition, and temperature are most commonly studied. The so-called critical molecular weight technique is useful when a change in polymer molecular weight causes a change in miscibility [12,13]. A recent review (1) summarizes values of B_{ij} for many repeat unit pairs of interest determined by this strategy and other methods such as small angle neutron scattering.

Theory of Polymer-Polymer Interfaces

As discussed above, binary interaction energies play a defining role in determining the phase behavior of polymer blends. These interactions are also important in phase separated systems since they strongly influence the nature of the interface between phases which, in turn, is a factor in determining morphology (via interfacial tension) in the melt state and adhesion (via interfacial thickness) in the solid state. A quantitative relation between the thermodynamic interaction energy and interfacial properties was first proposed by Helfand and Tagami [14]. The generalized theory formulated by Helfand and Sapse [15] for infinite molecular weight polymers describes the interfacial thickness, λ, as

$$\lambda = \sqrt{\frac{2RT}{B}} \left(\beta_A^2 + \beta_B^2\right)^{1/2} \qquad (6)$$

In this expression, B is the interaction energy density and β_i is related to the dimension of the polymer coil by

$$\beta_i = \sqrt{\frac{\rho_i}{6}} \left(\langle r_i^2 \rangle / M_i\right)^{1/2} \qquad (7)$$

where $\langle r_i^2 \rangle$ is the mean square unperturbed end-to-end chain distance and M_i is the molecular weight. Similarly, the interfacial tension, γ, is expressed in the same nomenclature as

$$\gamma = \sqrt{\frac{RTB}{2}} \left(\beta_A + \beta_B\right) \left[1 + \frac{1}{3} \frac{\left(\beta_A - \beta_B\right)^2}{\left(\beta_A + \beta_B\right)^2} \right] \qquad (8)$$

Equations 7 and 8 assume infinite molecular weights of both components; more recently, Broseta et al [16] extended the theory to treat finite molecular weights. For blend systems that include copolymers, the appropriate form of the binary interaction model, e.g., equation 5, can be used to evaluate B provided the appropriate B_{ij} values are known.

Comparison of Theory and Experiment

On the whole, predictions of λ and γ have been correlated with experimental measurements with varying degrees of success. Caution must be exercised in such comparisons, however, because the measurements are difficult and require extreme care in preparing the samples, executing the experiments, and interpreting the data. Furthermore, limitations of the theory must be recognized. When these issues are carefully considered, impressive agreement between theory and experiment has been found as illustrated below for two systems of some commercial interest both involving styrene/acrylonitrile copolymers (SAN) of varying AN content.

Figure 1 shows results for interfaces formed between bisphenol-A polycarbonate (PC) and
SAN copolymers. For this system, the PC/SAN interaction energy shows a minimum, but positive, value near the azeotropic composition for SAN copolymer, i.e., ~25% wt. AN [2]. Thus, the theory predicts that interfacial thickness is at a maximum at this composition while the interfacial tension is at a minimum. Mansfield [17,18] has reported

an experimental value for λ measured by neutron reflectivity near this composition that is in excellent agreement with the prediction of theory, see Figure 1a. Watson and Hobbs [19] measured the interfacial tension by a capillary thread instability method for PC and a series of SAN copolymers with the results shown in Figure 1b. The agreement with theory is generally good. These trends in λ and γ also correlate very well with observations on morphology, interfacial adhesion, and mechanical properties for this system as described by Callaghan et al [2].

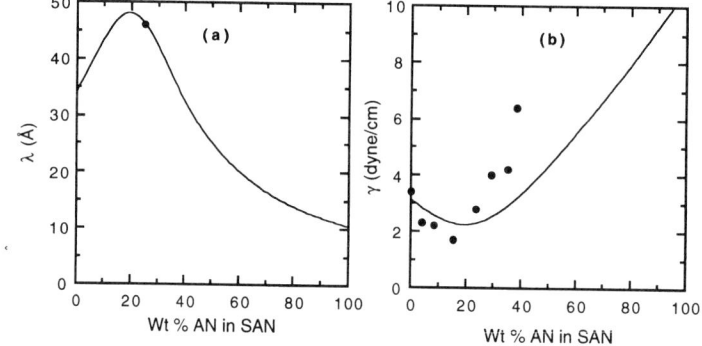

Fig. 1 Interfacial properties of PC and SAN as a function of copolymer composition (see reference 3). (a) Interfacial thickness calculated using equation 6 and binary interaction energies reported in the literature compared to a measurement made by neutron reflectivity (see reference 17). (b) Interfacial tension calculations based on equation 8 compared to measurements made by capillary thread instability (see reference 19). Calculations made at 140°C using $B_{S/PC} = 0.43$, $B_{PC/AN} = 4.5$, and $B_{S/AN} = 7.0$ cal/cm^3.

Figure 2 shows the interfacial thickness for PPO bilayers with SAN copolymers measured (solid points) by neutron reflectivity [3]. The solid curve is the theoretical prediction from thermodynamically measured values of B_{ij}. The agreement is quite good. For this system, PPO is miscible with polystyrene and miscibility is maintained until about 12% wt. of AN in the SAN copolymer. The PPO/SAN interaction energy is predicted to monotonically increase with AN content from negative values through the critical interaction energy and continuing to become more positive with higher AN levels. This correctly explains the trend in λ versus %AN observed by neutron reflectivity.

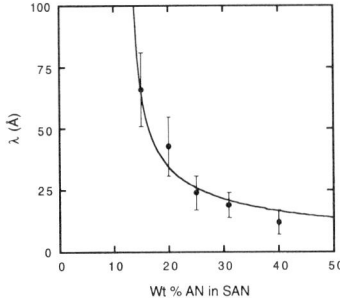

Fig. 2 Neutron reflectivity measurements and theoretical prediction of the PPO/SAN interfacial thickness as a function of SAN composition (see reference 3). The prediction is made using the interfacial theory formulated by Helfand and coworkers combined with the binary interaction model. The interface broadens exponentially as the AN content in the copolymer decreases toward the miscibility limit where SAN copolymers containing approximately 12 wt% and less re miscible with PPO. Calculations were made at 140°C using $B_{PPO/S}$=-0.34, $B_{PPO/AN}$=9.2, and $B_{S/AN}$=7.0 cal/cm^3.

Conclusion

The binary interaction approach provides a useful way of representing polymer-polymer interaction energies for copolymer systems. There is a growing body of literature on quantitative estimates of interaction energies between repeat unit pairs, B_{ij}, that can be utilized in the design and optimization of new polymer blends. This paper demonstrates that interfacial properties of immiscible blends can be predicted from this information using current theories of polymer-polymer interfaces and interactions.

Acknowledgment

Financial support for this work was provided by the National Science Foundation grant number DMR 97-26484 administered by the Division of Materials Research – Polymers Program.

References

1. G.D. Merfeld and D.R. Paul, Ch. 3 in "Polymer Blends: Formulation and Performance", D.R. Paul and C.B. Bucknall, editors, John Wiley, New York, 2000
2. T.A. Callaghan, K. Takakuwa, D.R. Paul, and A.R. Padwa, *Polymer*, **34**, 3796 (1993)
3. G.D. Merfeld, A. Karim, B. Majumdar, S.K. Satija, and D.R. Paul, *J. Polym. Sci.: Part B: Polym. Phys.*, **36**, 3115 (1998)
4. P. J. Flory, *J. Chem. Phys.*, **10**, 51 (1942)
5. M. L. Huggins, *J. Chem. Phys.*, **9**, 440 (1941)

6. P. J. Flory, *Principles of Polymer Chemistry,* Cornell University Press, Ithaca, New York, 1953
7. R. Koningsveld and H. A. G. Chermin, *Proc. R. Soc. London A,* **319**, 331 (1970)
8. R. Koningsveld and L. A. Kleintjens, *J. Polym. Sci., Polym. Symp.,* **61**, 221 (1977)
9. R. P. Kambour, J. T. Bendler and R. C. Bopp, *Macromolecules,* **16**, 753 (1983)
10. G. ten Brinke, F. E. Karasz and W. J. MacKnight, *Macromolecules,* **16**, 1827 (1983)
11. D. R. Paul and J. W. Barlow, *Polymer,* **25**, 487 (1984)
12. R. P. Kambour, P. E. Gundlach, I. C. W. Wang, D. W. White and G. N. Yeager, *Polym. Comm.,* **29**, 170 (1988)
13. T. A. Callaghan and D. R. Paul, *Macromolecules,* **26**, 2439 (1993)
14. E. Helfand and Y. Tagami, *J. Polym. Sci., Polym. Lett.,* **9**, 741 (1971)
15. E. Helfand and A. M. Sapse, *J. Chem. Phys.,* **62**, 1327 (1975)
16. D. Broseta, G. H. Fredrickson, E. Helfand and L. Leibler, *Macromolecules,* **23**, 132 (1990)
17. T. L. Mansfield, Ph.D. dissertation, University of Massachusetts (1993)
18. V. Janarthanan, R.S. Stein, and P.D. Garrett, *Macromolecules,* **27**, 4855 (1994)
19. V. H. Watkins and S. Y. Hobbs, *Polymer,* **34**, 3955 (1993)

Macromol. Symp. **159**, 113–121 (2000)

Surface Studies of Polyethers with Well-Defined Segmental Length

Chi-Ming Chan[*], Lin Li[*], Kai-Mo Ng[*], Jianxiong Li[*] and Lu-Tao Weng[**]
Department of Chemical Engineering[*]
Advanced Engineering Materials Facility[*]
Materials Characterization & Preparation Facility[**]
Hong Kong University of Science and Technology
Clear Water Bay, Kowloon, Hong Kong
P.R. of China

SUMMARY: Two series of polyethers were synthesized by the polymerization of 1, n-dibromoalkane (n = 4, 6, 8, 10, 12, 14 and 18) with bisphenol-A (BA) and 4, 4'-(hexafluoroisopropylidene) diphenol-A. The length of the flexible aliphatic segment changes from 4 Å to 21 Å (corresponding to 4 to 18 CH_2 groups). X-ray photoelectron spectroscopy (XPS), time-of-flight secondary ion mass spectrometry (ToF-SIMS) and atomic force microscopy (AFM) were used to characterize the surfaces of the polyethers. The influence of the length of the flexible aliphatic segments on the surface composition of the BA and 6FBA polyethers was investigated. The intensity ratios of the characteristic SIMS peaks of the flexible segments to those of the rigid segments were related to the length of the flexible segments and the XPS results confirmed that the SIMS intensity ratios can be used to determine the surface compositions. AFM was utilized to investigate the crystallization process of the BA polymer with n = 8. The appearance of nuclei, the generation of primary lamellae and the formation of spherulites were observed dynamically. Nuclei appeared first as 10 nm dots, some disappeared and a few could grow into lamellae. The lamellae that developed from the nuclei bred more lamellae, which in turn induced secondary nucleation and branching of lamellae and finally led to a spherical appearance.

Introduction

The surface segregation in polymers is governed by several factors. The thermodynamic driving force for minimizing the total free energy of the system results in preferential surface segregation of the lower surface energy constituent of the polymer. However, a number of other factors, such as the relative length of the blocks, their sequence length distribution and the concentration of the constituents may also play an important role in the surface composition of copolymers.[1-5] In one of our recent works, the effects of the sequence distribution of poly(acrylonitrile-butadiene) rubbers (NBRs), determined by ^{13}C-NMR, on the surface chemical composition, determined by XPS and dynamic contact angle (DCA) measurements, were studied.[5] The results of XPS and ^{13}C-NMR analyses suggest that surface segregation of the lower surface energy segment (BD) of the NBR copolymers occurs only when the BD segment length is at least equal to the length of the BBB segment and the BBB intensity is above a certain value.

          CCC 1022-1360/00/$ 17.50+.50/0

It has been shown that the secondary ion fragments emitted from a polymer surface are related to the surface molecular structure.[6,7] In particular, ToF-SIMS can provide structural information both in the positive and negative spectra. In order to investigate the effects of the sequence length on the surface composition of a polymer two series of polyethers with a well-defined segment length were prepared, the relationship between the structure of the secondary ion fragments and the surface chemical structure was studied.

Recently, AFM has been applied to study the morphology and kinetic phenomena of polymers.[8,9] In this work, the crystallization process of a semicrystalline polyether with a glass transition temperature near room temperature was studied.

Experiment

The synthesis is described as in Scheme 1:[10]

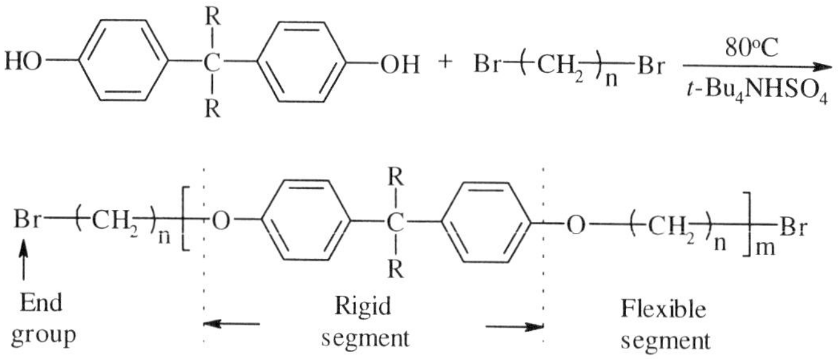

R: —CH₃ the BA polymers; —CF₃ the 6FBA polymers

n = 4, 6, 8, 10, 12, 14 or 18

Scheme 1: The preparation of the BA and 6FBA polyethers.

Thin films of the BA and 6FBA polymers were prepared by solution spin casting on silicon wafers. The solvents used for the BA and 6FBA polymers were chloroform and tetrahydrofuran, respectively.

XPS spectra were recorded on a PHI 5600 multi-technique system equipped with an Al monochromatic X-ray source. A pass energy of 58.7 eV was used. The spectra were obtained at a take-off angle of 45°. The ToF-SIMS experiments were performed on a Physical Electronics PHI 7200 ToF-SIMS spectrometer. AFM phase images were obtained using a DI NanoScope III MultiMode™ AFM at room temperature. Si tips with a resonance frequency of ~300kHz were used and the scan rate was 0.8 Hz sec⁻¹.

Results and Discussion

The influence of segmental length on the surface chemical composition

The atomic ratio of F-to-O is plotted as a function of the number of the CH_2 groups, n, as shown in Fig. 1[11]. This ratio is a constant, implying that there is no contamination on the surface of the 6FBA polymers. The concentration of the end groups, Br, is too low to be determined accurately in the BA and 6FBA polymers.

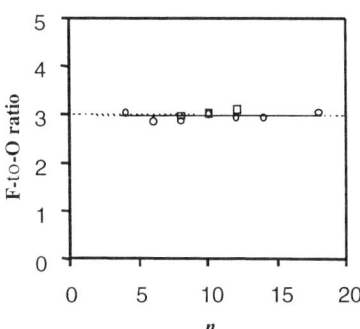

Fig. 1: The F-to-O atomic ratio *vs.* n of the 6FBA polymers

The number of the CH_2 groups, n_s, determined by the C-to-O atomic ratios of the BA and 6FBA polymers, as a function of n, is shown in Fig. 2. The value of n_s of the 6FBA polymers exhibits a linear relationship with the value of n for $n = 4$ to 18.

However, for the BA polymers n_s determined by XPS is much larger than n when $n > 14$. This suggests that the flexible segments segregate onto the surface of the BA polymers when the length of the flexible segment is long enough ($n > 14$, segment length ~17Å). Repeated measurements yielded similar results. A comparison of the XPS results between the BA and 6FBA polymers indicates that the effects of the flexible segment length on the surface composition are different even for the

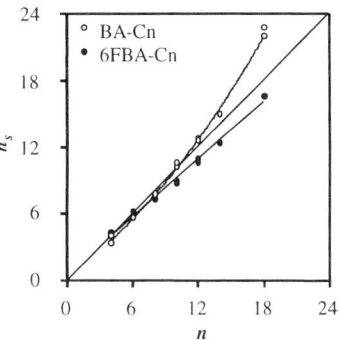

Fig. 2: n_s determined by XPS *vs.* n of the BA and 6FBA polymers

sequential polymers with a very similar structure. The difference in the surface free energy between the rigid and flexible segments is the driving force for the segregation of the low surface component to the surface. Because the BA and 6FBA polymers have a different side group (CH_3 and CF_3), the surface energy difference between the rigid and flexible segments is not identical for these two polymers. The length of the flexible segment plays an important role in determining the surface composition of the polyethers. In the BA polymers, the flexible segment obviously has a lower surface energy than the rigid segment. When the length of the flexible segment increases, the mobility of the segment also increases. As a result, surface excess of the flexible segments occurs. But for the 6FBA polymers, the surface energy difference between the rigid and flexible segments may not be significant. Therefore, no significant surface excess of the flexible segments is observed even at $n = 18$.

The influence of the segmental length on the SIMS intensities

The influence of the flexible segment length on the normalized intensity, I_i^N, of the positive aliphatic hydrocarbon ions and the aromatic ions was investigated. The normalized intensity of the characteristic fragments of the flexible aliphatic segments and rigid segments is defined by the following equation:

$$I_i^N = \frac{I_i / I_t}{I_i / I_t (n = 4)} \tag{1}$$

where I_t is the total ion intensity, $I_i / I_t (n = 4)$ is chosen as the reference and all normalized peak intensities of 6FBA polymers, $(I_i / I_t (n = 4)) = 1$. For example, I_i^N of the 6FBA polymers $vs.$ n is plotted in Fig. 3. A series of positive aliphatic ion fragments, such as $C_4H_2^+$, $C_4H_3^+$, $C_4H_4^+$, $C_4H_5^+$, $C_4H_6^+$, $C_4H_7^+$, $C_4H_8^+$ and $C_4H_9^+$, at m/z = 50, 51, 52, 53, 54, 55, 56, and 57 were used and the variation of I_i^N can be seen in Fig. 3a.

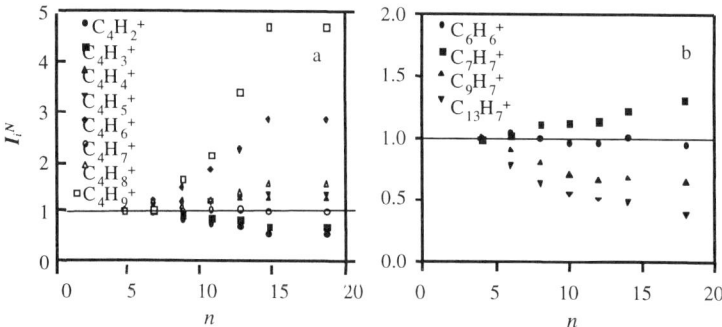

Fig. 3: The influence of the flexible segment length on the intensity of the characteristic positive hydrocarbon ions $vs.$ n of the 6FBA polymers. (a) aliphatic secondary ions and (b) aromatic secondary ions.

When the flexible segment length increases, the intensity of the highly unsaturated aliphatic ion fragments, such as $C_4H_2^+$ and $C_4H_3^+$ at m/z = 50 and 51 decreases. However, the intensity of the saturated aliphatic ion, $C_4H_9^+$, at m/z = 57 increases as the flexible segment length increases. This can be explained by the structure of the 6FBA polymers, as shown in Scheme 1. The highly unsaturated aliphatic ions are mainly emitted from the rigid segment and their intensity decreases as the number of the CH_2 group increases. The other unsaturated ions, such as $C_4H_4^+$, $C_4H_5^+$, $C_4H_6^+$, C_4H_7+ and $C_4H_8^+$, are emitted from both the rigid and flexible segments.

The variation of I_i^N of the aromatic ion fragments of the 6FBA polymers, such as $C_6H_5^+$, $C_7H_7^+$, $C_9H_7^+$ and $C_{13}H_9^+$ at m/z = 77, 91, 115 and 165, is shown in Figure 3b. The aromatic ions such as $C_6H_5^+$ and $C_7H_7^+$ can also be emitted from the aliphatic flexible segments. Therefore, they cannot be used as the characteristic positive ions of the rigid segments. However, the larger aromatic fragments, such as $C_9H_7^+$ and $C_{13}H_9^+$, are mainly emitted from the rigid segments; therefore, it is possible to use the intensity of these ions to determine the surface composition of the 6FBA polymers.

In order to simplify the analysis of the ToF-SIMS results, the characteristic ion fragments of the BA and 6FBA polymers are summarized in three series. (1) Flexible fragments, F_i, are mainly emitted from the flexible segment, such as CH_2^-, $C_2H_5^+$, $C_3H_7^+$, and $C_4H_9^+$, at m/z = 14, 29, 43, and 57, respectively. (2) Rigid fragments, R_j, are mainly emitted from the rigid segment, such as $C_7H_{11}O^+$ and $C_{14}H_{13}O_2^+$ at m/z = 135, and 213, respectively, for the BA polymers and $C_9F_6H_5O^+$, $C_{14}F_3H_{10}O_2^+$, $C_9F_5H_4O^-$ and $C_{14}F_3H_8O_2^-$ at m/z=243, 267, 223 and 265, respectively, for the 6FBA polymers. The chemical structures of these ion fragments were reported previously[10]. (3) Molecular fragments, AB_n, are directly related to the structure of the repeat units of the BA and 6FBA polymers, such as [M-H]$^-$ at m/z = 281, 309, 337, ... with n = 4, 6, 8, ... of the BA polymers. If the bulk effects can be neglected, the ratios of the absolute intensities, $I(F_i)/I(R_j)$, of the flexible fragments to the rigid fragments are related to the number of the CH_2 groups at the surface, n_s, as shown in Eq. 2:

$$\frac{I(F_i)}{I(R_j)} = \frac{S_i \cdot C(F)}{S_j \cdot C(R)} \propto K(ij) \cdot n_s \tag{2}$$

where $I(F_i)$ and $I(R_j)$ are the absolute intensities of the flexible and rigid segments, respectively. S_i and S_j are related to the sensitive factors of the ion fragments i and j, respectively. $C(R)$ and $C(F)$ are the molar concentrations of the flexible and rigid segments, respectively. K_{ij} is a parameter related to the yield of fragments i and j.

To verify Eq. 2, the SIMS results of the BA and 6FBA polymers were analyzed. The intensities of the positive saturated aliphatic ions, such as $C_2H_5^+$, $C_3H_7^+$ and $C_4H_9^+$ at m/z = 29, 43, and 57, $I(F_{29})$, $I(F_{43})$, and $I(F_{57})$, were used for the flexible segments of the polymers. For the rigid segments of the BA polymers, the intensities of the characteristic positive ions of the BA polymers $C_9H_{11}O^+$ and $C_{14}H_{13}O^+$ at m/z =135 and 213, $I(R_{135})$ and $I(R_{213})$, were adopted, while for the 6FBA polymers the intensities of the characteristic positive ions of the 6FBA polymers $C_9F_6H_5O^+$ and $C_{14}F_3H_{10}O_2^+$ at m/z =243 and 267, $I(R_{243})$ and $I(R_{267})$, were adopted. As an example, the intensity ratios of the ions for the BA polymers, $I(F_{57})/I(R_{135})$ $(C_4H_9^+/C_9H_{11}O^+)$ and $I(F_{57})/I(R_{213})$ $(C_4H_9^+/C_{14}H_{13}O_2^+)$, are plotted *vs.* n, as shown in Fig. 4. The intensity ratios, $I(F_{57})/I(R_{135})$ and $I(F_{57})/I(R_{213})$, increase monotonically with n, as shown in Fig. 4a. It is obvious that the ratios do not increase linearly with n. However, the intensity ratios, $I(F_{57})/I(R_{135})$ and $I(F_{57})/I(R_{213})$, increase fairly linearly with n_s as shown in Fig. 4b. The intensity ratios of 6FBA polymers, $I(F_{57})/I(R_{243})$ $(C_4H_9^+/C_9F_6H_5O^+)$, and $I(F_{57})/I(R_{267})$ $(C_4H_9^+/C_{14}F_3H_{10}O_2^+)$, increase linearly with n. The other intensity ratios,

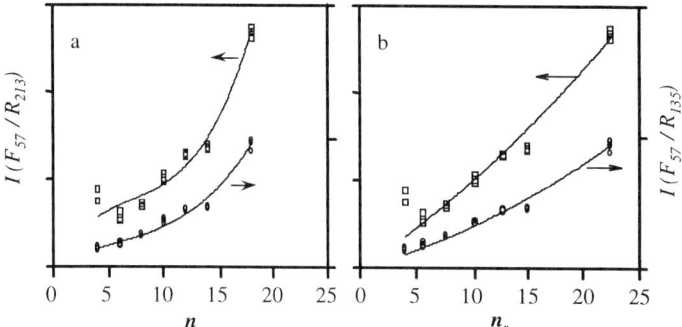

Fig. 4: Intensity ratios of positive ions $I(F_{57})/I(R_{135})$ and $I(F_{57})/I(R_{213})$ for the BA polymers as function of (a) n and (b) n_s.

$I(F_{29})/I(R_j)$ and $I(F_{43})/I(R_j)$, show similar results. These results further reflect the fact that the ratios of the ionic intensities are directly related to the surface concentration of the flexible segments and can be used for quantitative analyzing the surface compositions of these polymers.

Nuclei formation and spherulite growth

AFM was utilized to study the crystallization process and the organization of the spherulite of the BA polymer with n=8.[12] This polymer is attractive for AFM analyses because its glass transition temperature is close to room temperature and its crystallization rate at room temperature allows imaging of the crystallization process with an AFM without a hot stage. The glass transition temperature, melting point, and number–average molecular weight of this polymer were measured to be 6.9°C, 83.5°C, and $5.7×10^3$ g/mole respectively. A thin film about 300 nm was prepared by spin casting the polymer solution

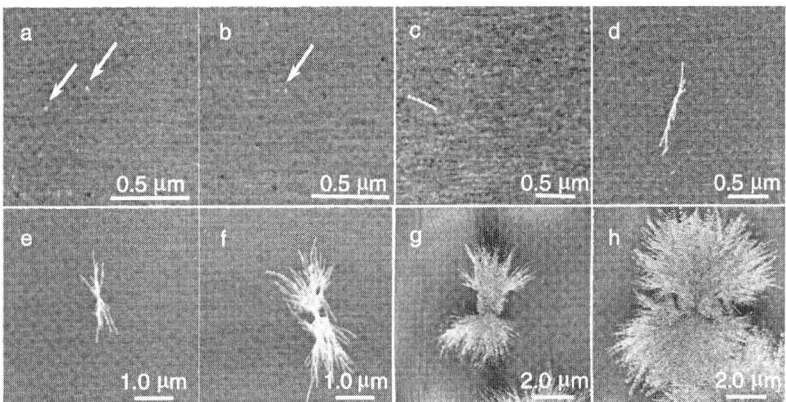

Fig. 5: The formation of a primary nucleus, lamellae and a spherulite

on a silicon wafer and the crystallization process was observed directly under tapping-mode AFM phase imaging, as shown in Fig. 5. Embryos of the lamellae, as fine dots, with a dimension smaller than 10 nm, appeared and disappeared on the surface of the amorphous polymer film during scanning (Figs. 5a and 5b). Figure 5a shows the presence of two embryos. In Fig. 5b, which was obtained approximately 8.7 minutes after the image shown in Fig. 4a, one of the embryos disappeared. We believe that this is the first experimental evidence in a polymer to show that, as predicted by thermodynamics, embryos smaller than a critical dimension are unstable and may not ultimately grow. The initial nucleating lamella grows along the length of the lamella at both ends (Fig. 5c). During the growth stage, it breeds more lamellae (Fig. 5d). When their lengths are longer than 0.5 to 1.0 μm, the lamellae begin to form branches (Fig. 5e) and some lamellae splay apart from each other (Fig. 5f). As a result of continual splaying and branching of the lamellae, the initial lamellae gradually develop into a lamella sheaf and a spherulite skeleton (Figs. 5g, h). We believe that all lamellae except for those in the "eye" of the spherulites are viewed edge-on because the "width" of the lamellae, as viewed in this direction, is approximately 10 nm and this dimension is likely to be the thickness of the lamellae.

It has been suggested that a polymer spherulite develops from a single lamella through unidirectional growth and continuous branching and splaying apart [9, 13]. Our examination on the development of a lamella embryo into a mature spherulite provides direct evidence to support this view. Furthermore, our experiments revealed that the periphery of the developing spherulites is not as smooth as that observed under optical microscopy. The growth front of the spherulites looks like a hedgehog, as shown in Fig. 5h.

Conclusion

The XPS results indicate that for the low energy segments to segregate on the surface of the polyethers, the segmental length of the low surface energy component has to be above a certain value. The SIMS intensity ratios of the ions from the flexible segments and the rigid segments are related to the surface concentration of the flexible segments as determined by XPS. These findings demonstrate that SIMS data can be used for quantitative analyses of polymer surface. The AFM results provide the direct experimental evidence to show that the embryos of crystalline nuclei are unstable when their dimensions are smaller than a critical value and not all the embryos can ultimately develop into lamellae.

References

1. M. J. Hearn, B. D. Ratner and D. Briggs, *Macromolecules,* **21**, 2950 (1988)
2. M. J. Hearn, D. Briggs, S. C. Yoon, B. D. Ratner, *Surf. Interface Anal.,* **10**, 384 (1987)
3. P. L. Kulmer, H. L. Matteson and Jr., J. A. Gardella, *Langmuir*, **7**, 2479 (1991)
4. X. Chen, Jr., J. A. Gardella, T. Ho and K. J. Wynne, *Macromolecules,* **28**, 1635 (1995)
5. L. Li, C. M Chan and L. T. Weng, *Macromolecules,* **30**, 3698 (1997)

6.   D. Briggs, in *Surface Analysis of Polymers by XPS and Static SIMS,* Cambridge University Press, 1998; L. Li, C. M Chan and L. T. Weng, *Macromolecules,* **30**, 3698 (1997)

7.   C. M. Chan, in *Polymer Surface Modification and Characterization,* Hanser, New York, 1994

8.   R. Pearce and G. J. Vancso, *Polymer*, **39**, 1237 (1998)

9.   D. A. Ivanov and A. M. Jonas, *Macromolecules*, **31**, 4546 (1998)

10.   L. Li, C. M. Chan, K. M. Ng and L. T. Weng, *Macromolecules* (submitted).

11.   L. Li, C. M. Chan, S. Liu, L. An, K. M. Ng, L. T. Weng and K. C. Ho, *Macromolecules* (submitted)

12.   L. Li, C. M. Chan, J. X. Li, K. M. Ng, K. L. Yeung and L. T. Weng, *Macromolecules* **32,** 8240 (1999)

13.   R. H. Olley and D. C. Bassett, *Polymer* **30**, 399 (1989)

Macromol. Symp. 159, 123–130 (2000)

Effect of Dicarboxy Terminated Polystyrene on Strengthening Immiscible Polystyrene/ Poly(methyl methacrylate) Interface

Kookheon Char*, Yeonsoo Lee, Yooseong Yang, and Changhoon Sim

School of Chemical Engineering, Seoul National University, 56-1 Shinlim-dong, Kwanak-ku, Seoul 151-742, KOREA

SUMMARY: The fracture toughness between polystyrene (PS)/ poly(methyl methacrylate) (PMMA) reinforced with reactive polymers, poly(glycidyl methacrylate) (PGMA) and *dicarboxy or monocarboxy* terminated PS (dcPS and mcPS), was measured by the asymmetric fracture test. Molecular weight effect of mcPS, although the molecular weight distribution is rather polydisperse, on the maximum achievable fracture toughness, G_{max} qualitatively agreed with the results of the monodisperse case[4,5]. In the case of dcPS with $M_w \cong 142$ K, G_{max} reached ca. 170 J/m^2 which is nearly 8 times higher than that of mcPS of molecular weight of about 150K. From the mechanical point of view, dcPS with a degree of polymerization (N) greater than the ratio of chain breaking force to monomeric friction force (f_b/f_{mono}) is more effective in enhancing the interfacial adhesion than mcPS since it provides two *stitches* to the interface. It was also shown by Monte Carlo simulation on reactive polymer system that the di-endfunctional polymers are more effective than mono-endfunctional polymers in reinforcing the week interface between immiscible polymers.

Introduction

One of major goals of using compatibilizers in polymer blends or alloys is to enhance the otherwise weak interfacial adhesion between immiscible polymers. Block copolymers, random copolymers and reactive polymers have been reported as strong candidates to achieve this goal since they provide stress transferring *bridges* or *stitches* to the weak interface by being preferentially located at the interface[1-5].

Recently Kramer and coworkers[4,5] demonstrated that in the study of the fracture behavior of polystyrene (PS)/epoxy interface reinforced with carboxy terminated deuterated PS (d-

CCC 1022-1360/00/$ 17.50+.50/0

PSCOOH) having narrow molecular weight distribution which was used as a model reactive polymer there exists an optimum molecular weight of d-PSCOOH in enhancing the interfacial adhesion which is balanced by the two competing effects: the molecular weight dependency of the nature of craze against the molecular weight effect of the saturation chain areal density (Σ_{sat}) of d-PSCOOH. The fracture toughness showed the maximum value at a molecular weight about $N \sim f_b/f_{mono}$ which corresponds to $N \sim 1000$ in the case of d-PSCOOH, where f_b is the force to break a chain and f_{mono} is the monomeric friction force. However, the maximum achievable areal density (Σ_{max}) decreased almost linearly with respect to molecular weight ($\sim 1/N$) in the experiment, which is in contrast to $1/N^{0.5}$ in the case of a Gaussian graft chain. To explain this linear decrease they took into account the entropic barrier that resists the addition of new monocarboxy terminated chains to the end grafted chains already formed by the reaction at interface as well as the decrease of Σ_{sat} with increasing the molecular weight of d-PSCOOH[5].

In this study, based on the fracture map mentioned above, we focus on the effect of *dicarboxy* terminated PS (dcPS) containing two carboxyl groups at both ends of a PS chain on the fracture behavior of immiscible polymer interface in comparison with the case with monocarboxy terminated PS (mcPS). PS / poly(methyl methacrylate) (PMMA) were chosen as an immiscible weak polymer interface. poly(glycidyl methacrylate) (PGMA), which is compatible with PMMA, was employed as a reactive polymer. PGMA has inherently epoxy functional groups reacting readily with carboxyl group of carboxy terminated PS (cPS) which was employed as the counterpart reactive polymer. We also performed the Monte Carlo simulations on the mono- and di-endfunctional polymer reaction system and qualitatively compared with the experimental results obtained from the fracture test.

Experimental Part

Materials and measurement methods
Commercial grade PS (from Miwon Petrochemical Co.) and PMMA (from LG Chemical) were dried in a vacuum oven and used to prepare bulk homopolymer sheets by compression molding. PGMA and cPS were synthesized using free radical polymerization. To place a

carboxyl group at the end of a PS chain a small amount of mercapto acetic acid was added as a chain transfer agent[6]. dcPS used in this study was purchased from Aldrich. Details of the reactive polymers used in present study are summarized in Table 1.

Table 1. Reactive polymers used in this study.

Code	M_n	M_w	Source
mcPS29	12 K	29 K	Synthesized
mcPS45	15 K	45 K	Synthesized
mcPS60	34 K	60 K	Synthesized
mcPS150	46 K	147 K	Synthesized
mcPS170	50 K	165 K	Synthesized
mcPS235	120 K	235 K	Aldrich
dcPS142	93 K	142 K	Aldrich
PGMA	230 K	500 K	Synthesized

To prepare specimens for the interfacial adhesion test, PGMA dissolved in methylethylketone (MEK) was spun onto the PMMA plate. The film thickness of PGMA was fixed at 1,500 nm for present study. The counterpart reactive polymer, cPS, was dissolved in toluene. The cPS films with thicknesses ranging from 15 nm to 1500 nm was first spun on glass slides, floated on distilled deionized water and then transferred to PS homopolymer sheet. The PS and PMMA sheets containing reactive polymers on each sheet were then joined together at 190 °C for 1 hour in a hot press. The fracture toughness of interface was measured by the asymmetric fracture test[7].

Results and discussion

The results of the fracture toughness between PS and PMMA measured as a function of cPS thickness are given in Fig. 1. The fracture toughness of PS/PMMA initially increases with increasing cPS thickness and then levels off at a constant value. This implies that the amount of copolymer formed *in-situ* at the PS/PMMA interface increases up to a plateau value and further increase in the amount of copolymer at the interface is not possible even though cPS thickness is increased up to 1,500 nm. The effect of the initial amount of the reactive cPS polymers on the interfacial adhesion can be explained by the result of the competition between the reaction at interface and the dilution of cPS concentration at the interface due to

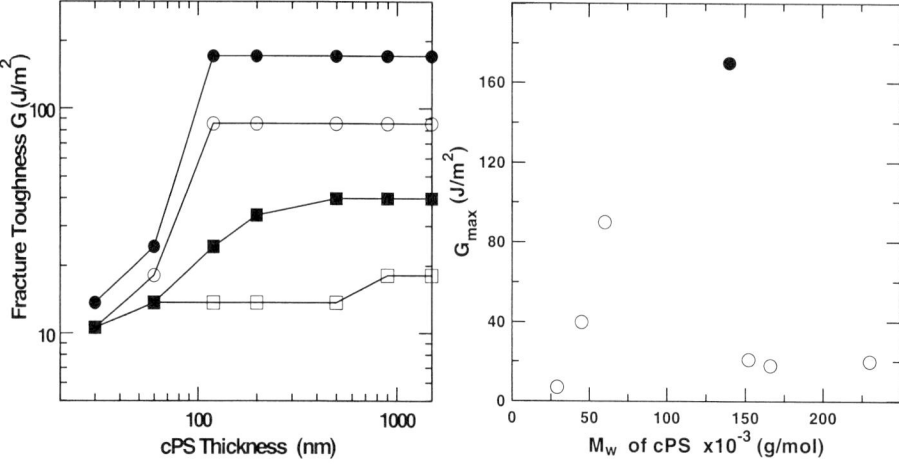

Fig. 1: Fracture toughness between PS/ PMMA plotted against initial thickness cPS: (●) dcPS142, (○) mcPS60, (■) mcps45, (□) mcps235.

Fig. 2: The maximum achievable fracture toughness as a function of M_w of cPS: (○) mcPS, (●) dcPS.

the diffusive mixing of the reactive polymers with bulk homopolymer. Detailed results of the effect of cPS thickness on the fracture toughness of PS/PMMA will be given elsewhere[8]. In this paper, our main focus is placed on the maximum achievable fracture toughness value for a given molecular weight of cPS.

As shown in Fig. 2, the maximum fracture toughnesses was plotted against the weight average molecular weight (M_w) of cPS. Since cPS used in this study is polydisperse in molecular weight distribution, the number average molecular weight (M_n) instead of M_w may be an alternative to plot the fracture data. But in comparison of the data for mcPS29 with those for mcPS45 the difference in M_w (16 K) describes the difference in fracture toughness better than that in M_n (3 K) and thus M_w is chosen in this study as a molecular weight variable. In case of mcPS, if the polydispersity of cPS is taken into account, the fracture behavior in Fig. 2 qualitatively agrees with the previous results with monodisperse mcPS[4,5]. It is quite interesting to note the different fracture toughness result with dcPS. The fracture toughness of dcPS142 yields the highest value among all cPS's used in this study as shown in Fig. 2. The fracture toughness reaches ca. 170 J/m^2 which is about 8 times higher than that for mcPS150.

To better understand the fracture behavior, we indirectly estimated the maximum areal densities at the interface (Σ_{max}) by using the measured maximum fracture toughness values and eq 1 given below:

$$G = 2\pi\left(1 - \tfrac{1}{\lambda}\right)\sigma_c D\left[-\alpha\ln\left(1 - \left[\sigma_c / f_b \Sigma_{eff}\right]^2\right)\right]^{-1} \tag{1}$$

$$\Sigma_{eff} = q\Sigma\left(1 - \left(M_e / qM_n\right)\right)$$

Eq. 1, which was successfully developed by others[9,10], relates the macroscopic fracture toughness, G, to the microscopic chain areal density, Σ. The material properties of PS related to the microstructures of crazes formed at the crack tip are found as follows[10,11]: α (parameter related to the angle between main fibril and cross tie fibril) = 0.055, λ (craze fibril extension ratio) = 4, σ_c (craze stress) = 3.5 x 10^7 N/m^2, f$_b$ (force to break a chain) = 2 x 10^{-9} N, D (craze fibril diameter) = 8.65 nm, q (probability that a chain is in the craze after crazing) = 0.6, and M$_e$ (entanglement molecular weight) = 18,000 g/mol. The estimated maximum areal densities (Σ_{max}) are given in Fig. 3. Fig. 3 also shows, for comparison, the reported best fit of the Kramer group's data[4]. While the estimated Σ_{max} of mcPS in this study is almost the same as that of d-PSCOOH for the PS/epoxy interface, Σ_{max} for dcPS is about twice the Σ_{max} for the same molecular weight of mcPS. Considering the possible structure of the resulting copolymer formed, mcPS similar to block copolymer provides only one stress transferring *stitch or bridge* to the weak interface while dcPS provides two *stitches* to the interface. In the region where cPS molecular weight is greater than 100 K (N ~ f$_b$/f$_{mono}$), one reason that causes the decrease in the fracture toughness is the decrease in the saturation areal density with increasing the cPS molecular weight which in turn results in the decrease of the number of stitches per interfacial area. This effect is dominant over the strong craze formed in this molecular weight region[5]. Consequently, the increase in the number of stitches per chain such as the case for dcPS has a significant effect on the increase of interfacial adhesion from the mechanical point of view.

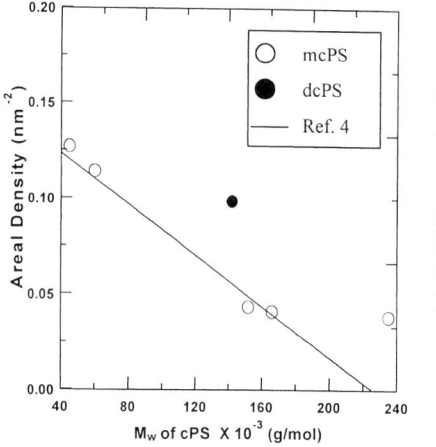

Fig. 3: The estimated Σ_{max} as a function of M_w of cPS: ($\bigcirc$) mcPS, ($\bullet$) dcPS. The line is the reported best fit of Kramer group's data[4].

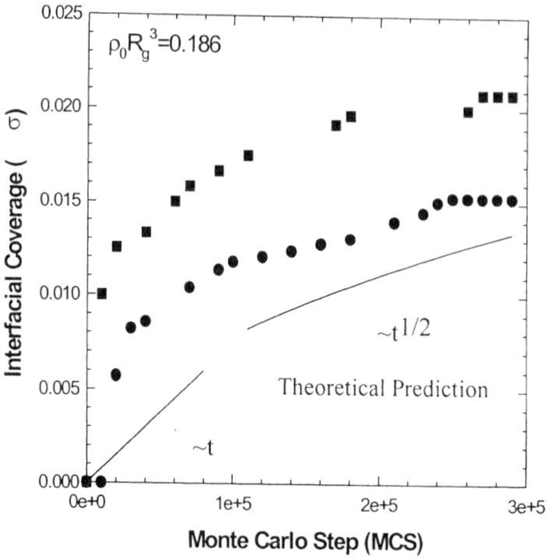

Fig. 4: Time dependence of copolymers formed at the interface due to reaction for $\rho_0 R_g^3 = 0.186$: ($\bullet$) mono- and ($\blacksquare$) di-endfunctional polymer reaction system, respectively.

In order to investigate the effectiveness of multiple reaction sites in detail, the reactions at immiscible polymer interface are examined by using the Monte Carlo (MC) simulation technique[12-14]. MC simulation results provide the information on time-dependence of the copolymer formation at the interface. Fig. 4 shows one example of the MC simlations with $\rho_0 R_g^3 = 0.186$, where ρ_0 is the number density of reactive polymers and R_g is the radius of gyration.

In Fig. 4 the early reaction rate and the plateau value of the interfacial copolymer coverage for di-endfunctional polymer reaction system are much larger than those for mono-endfunctional polymer reaction system. In the case of di-endfunctional polymer reaction system, there are two possibilities for the conformation of the copolymer formed due to reaction at the interface: loops and tails. For the case of the di-endfunctional reactive polymers, we investigated the distribution of loop and tail conformation as shown in Table 2. Loop conformations are dominant over the tail conformation. This implies that various polymer conformations, such as loops and tails, obtained with the di-endfunctional reactive polymers support the effectiveness of multiple reaction sites in enhancing interfacial properties like interfacial width and fracture toughness as mentioned above.

Table 2. Distribution of loop and tail conformation of copolymers formed at interfaces for di-endfunctional polymer reaction system with $\rho_0 R_g^3 = 0.557$.

Di-endfunctional Polymer Conformation	Loop	Tail
%	89.4	6.7

Conclusion

There exists a molecular weight range of cPS ($N > f_b/f_{mono}$) where dcPS is more efficient than mcPS in increasing the interfacial strength of immiscible polymers since it has potential to provide more than one *stitch* per molecule from a mechanical point of view and the formation of loop structure for dcPS which is necessary to provide two stitches per chain to the interface is possible from a thermodynamic point of view.

Monte Carlo simulation results also demonstrated that the di-endfunctional reactive polymer has advantage over the mono-endfunctional polymer in terms of the increase in the initial reaction rate and the amount of copolymers necessary to reinforce the weak interface.

Acknowledgment

This work was supported by the Korea Science and Engineering Foundation (KOSEF) under Grant 94-0520-02-3. We are very grateful to the financial support from the Brain Korea 21 Program through the Ministry of Education of Korea.

References

1. H. R. Brown, V. R. Deline, P. F. Green, *Nature(London)* **341**, 221 (1989)
2. C.–A. Dai, B. J. Dair, K. H. Dai, C. K. Ober, E. J. Kramer, C.–Y. Hui, L. W. Jelinski, *Phys. Rev. Lett.* **73**, 2472 (1994)
3. Y. Lee, K. Char, *Macromolecules* **27**, 2603 (1994)
4. L. J. Norton, V. Smigolova, M. U. Pralle, A. Hubenko, K. H. Dai, E. J. Kramer, S. Hahn, C. Berglund, B. Dekoven, *Macromolecules* **28**, 1999 (1995)
5. E. J. Kramer, L. J. Norton, C.–A. Dai, Y. Sha, C.-Y. Hui, *Faraday Discuss.* **98**, 31 (1994)
6. Y. Tsukahara, Y. Nakanishi, Y. Yamashita, H. Ohtani, Y. Nakashima, Y. F. Luo, T. Ando, S. Tsuge, *Macromolecules* **24**, 2493 (1991)
7. H. R. Brown, *J. Mater. Sci.* **25**, 2791 (1990)
8. Y. Lee, K. Char, *Macromolecules* **31**, 7091 (1998)
9. H. R. Brown, *Macromolecules* **24**, 2757 (1991)
10. Y. Sha, C. –Y. Hui, A. Ruina, E. J. Kramer, *Macromolecules* **28**, 2450 (1995)
11. E. J. Kramer, L. L. Berger, *Adv. Polym. Sci.* **91**, 1 (1990)
12. I. Carmesin, K. Kremer, *Macromolecules* **21**, 2819 (1988)
13. H. P. Deutsch, K. Binder, *J. Chem. Phys.* **94**, 2294 (1991)
14. M. Müller, *Macromolecules* **29**, 6353 (1997)

Surface Improvements of Aramid Fibers
By Physical Treatments

Hirosuke Watanabe[*1], Masashi Furukawa[*1], Tadahiko Takata[*2]
and Masahide Yamamoto[*3]

*1: Industrial Material Development Center, Teijin Ltd., Japan

*2: Teijin Cord (Thailand) Ltd., Thailand

*3: Ritsumeikan University (Emeritus Professor of Kyoto University), Japan

SUMMARY: The application of the low-temperature plasma method, the excimer laser treatment method and the corona-discharge method to aramid) were discussed, presenting an overview of current trends and developments in this area.

Introduction

New applications and improved applicability of the many fibers used for industrial materials require the new properties in such area as adhesive ability and the additional properties increase the value of fibers. Chemical treatment methods are most often used in actual practice. Physical treatment processes are dry, and make it possible to preserve certain properties intrinsic to fibers. Since 1960 this has been extensively studied.

The studies initially focused on electron beam irradiation and ultraviolet light irradiation, but electron beam irradiation required too much energy. Chemical treatment methods make use of large amounts of water and/or organic solvents, require high thermal energy for drying and curing, and pose such problems as the disposal of drained water and recovery of organic solvents. Therefore, it may be assumed that studies aimed at the eventual implementation of physical treatment methods will continue.

Para-type aramid fibers (abbreviated as aramid fibers) are useful as reinforcing fibers for matrices of rubber, resin etc.. However, the surfaces of the aramid fibers are relatively inactive, and their adhesive properties to rubbers, resins are insufficient. In order to enhance the adhesive properties for the rubbers and resins and attain the highest quality composite materials, the activation of the aramid fiber surfaces is extremely important.

The applications of the low-temperature plasma method and the excimer laser irradiation method to aramid fibers were discussed.

Experimental

Materials

The aramid fibers used in this study were PPODPTA yarn, made by Teijin Ltd. and PPTA yarn, made by Toray-DuPont Inc. The chemical structures are shown in Fig.1.

PPODPTA;

co-polyparaphenylene/3,4'-diaminodiphenylene terephthalamide

PPTA;

Polyparaphenylene terephthalamide

$$-\left(HN-\bigcirc-NHOC-\bigcirc-CO\right)_m\left(HN-\bigcirc-O \bigcirc -NHOC-\bigcirc-CO\right)_n$$

$$-\left(HN-\bigcirc-NHOC-\bigcirc-CO\right)_n$$

Fig.1 Chemical structures of aramid fibers

These were 1000 denier drawn yarn, and a plain weave fabric woven at 31 threads per inch for both warp and weft was applied. Scouring and de-oiling the fabric were done in the usual way. Table 1 gives the physical properties. These fibers are characterized by high tenacity and high degree of orientation and crystallinity. This was confirmed with wide-angle X-ray diffraction measurements.

Table 1.   Physical properties of aramid fibers

	Density	Diameter	Young's modulus	Tensile strength
PPODPTA	1390 Kgm^{-3}	12.4 μm	73.5GPa	3.04 GPa
PPTA(Type 29)	1440	12.1	58.8	2.79
PPTA(Type 49)	1450	12.0	127.4	2.79

Low Temperature Plasma Treatment

The aramid fabric samples were treated for predetermined time with the RFIP (radio frequency ion-plating) apparatus, as shown, at frequency of 13.56 MHz, at a discharge output of 200 W and at a vacuum of 10^{-3} torr using oxygen or nitrogen gases.

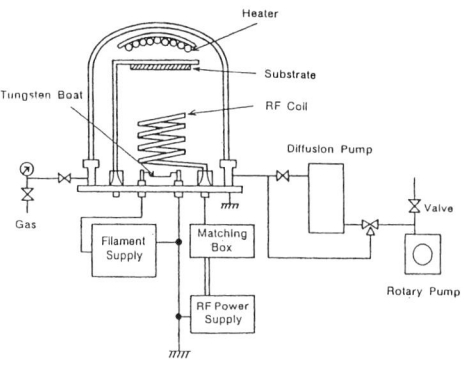

Fig. 1 Apparatus for Plasma Treatment

On the other hand, for RFIP, the aramid fabric samples were previously treated with low temperature plasma for 5 minutes under the same conditions as described above, using oxygen or nitrogen gas.

Different types of nylon were then melted at a current of 100A by the resistance-heating method, method, the shutter was opened, and oxygen or nitrogen gas was used as the carrier for a predetermined time, so that the different types of nylon were ion-plated on the aramid woven fabrics.

This treatment is based on the action of different active seeds, which results in etching (the developing of micro-irregularities) on the surface of fibers, the introduction of functional groups, and bridge building on the surface. However, there are problems such that the quality of the treated surfaces possibly degrades with times[1]. Therefore, it is said that the low temperature plasma treatment is not a stable method.

In this study, in order to improve the adhesive properties of aramid fibers to rubber as a matrix, nylon thin films were securely formed on the surface of the aramid fibers by RFIP technique, which was one application of the low temperature plasma treatment. The fibers were treated with a RFL (resorcin-formalin-latex) adhesive having a high affinity to nylon and rubbers. The adhesive properties of the fibers to rubber, were evaluated, and the effects of the treatment were confirmed by comparison with those obtained by the low temperature plasma treatment. The usefulness of the surface modification as a novel fiber surface activation method was discussed.

Excimer Laser Irradiation Method

This method can be applied in a normal atmosphere. Irradiation was carried out with a Lambda-Physik/Germany excimer laser apparatus filled with a mixture of krypton, neon and fluorine gases (wavelength 248nm). Each pulse of 20ns duration, delivered irradiation of 40, 50 or 100mJ/cm^2 above the threshold value.

In order to clarify the effects of the surface improvement on the adhesion performance with rubber, the following procedure was carried out. After being dipped in to an epoxy compound solution (5 wt-%), a sample of plain weave fabric was irradiated immediately with KrF excimer laser or was heated in an oven at 200℃. This was followed by dipping the sample in a RFL rubber adhesive and curing with heat (200℃ for 4 min).

To compare the adhesive performance, an alternative procedure was also applied. After being irradiated with a KrF laser, the sample was dipped into an epoxy compound solution and then cured in an oven at 200℃. This was followed by dipping in a RFL adhesive and curing with heat(200℃, for 4 min).

Estimation of Adhesion Performance

Test piece for adhesion strength evaluation were prepared by sandwiching a blended rubber (containing natural rubber/SBR) sheet, between two fabric samples placing the press in a mold and vulcanizing in a hot press at 150℃ for 30 min (for original adhesion) and at 180℃ for 60 min (for over-cured adhesion). An adhesive strength was measured by peeling (1 inch width) at a speed of 100 mm/min using a Model 5565 tensile tester made by INSTRON Corp/England.

Results and Discussion

Low Temperature Plasma Treatment

Changes in Fiber Surface

The changes in the aramid fibers at the surface[2], in the case of PPODPTA and PPTA, the oxygen gas low temperature plasma treatment produced so-called "Seashore Structure" perpendicular to the fiber axial direction. It was presumed that these convex and concave structures might be attributed to etching. The XPS analysis showed that the ratio of oxygen in the case of oxygen plasma was increased, thus indicating activation. It should be noted that such surface changes are not always observed, and depend on the type of gas.

Adhesive Properties

For the purpose of improving the adhesive properties, few reports on improvement of the adhesive properties to rubbers were found[3]. In the case of the low temperature plasma treatment using oxygen and nitrogen gases are shown in Fig. 3. These gases exhibited almost the same effect. By over-vulcanization, the aramid fibers exhibited less reduction or in some cases, an increase in adhesive properties, which is different from PET fibers of which the adhesive properties are reduced by over-vulcanization.

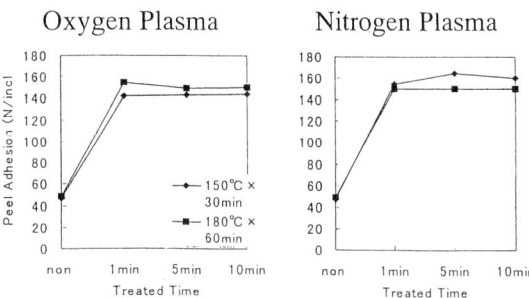

Fig.3 Adhesive properties to rubber

This is because the surfaces of the PET fibers deteriorated due to water and amine type vulcanizing agents contained in the rubber when the fibers were over-vulcanized. On the other hand, these behaviors of the aramid fibers are probably because they had less heat shrinkage and less degradation with heat generated during the vulcanization or rather had the interfaces strengthened by the anchoring effect due to good penetration of the rubber among the fibers by heating.

In comparison, PPODPTA woven fabrics had higher adhesive properties than the PPTA woven fabrics. The reason, while not completely understood, may lie in the fact that PPODPTA has a chemical structure more ready to be activated by the low temperature plasma treatment as compared with PPTA. In the case of PPTA, fibrils are more readily formed as compared with PPODPTA, which is a copolymer, and also weak boundary layers are produced by the low temperature plasma treatment.

Adhesive Properties of Nylon Coated Aramid Fibers

Thin Film Formation Condition
Till now, studies of thin film formation by the RFIP method have been made for addition polymerization type polymers such as polyethylene, polydiacetylene, polyvinylidene fluoride, etc.[4]. On the other hand, few reports of thin film formation using condensations type polymers such as polyesters, nylons, etc. were found. This is probably because such polymers may be pyrolyzed when they are heated to melting and then vaporized under vacuum conditions. Nylon 6 chips placed in a tungsten board in the RFIP were melted with the resistance-heating system, with the plasma atmosphere generated by flowing-in oxygen or nitrogen gas at a vacuum of $10^{-4} \sim 10^{-3}$ torr, and then discharged of 150 W. The shutter was opened, and the melted nylon 6 was vaporized and was deposited on an aluminum foil substrate (as a model test). The aluminum foil was dissolved in alkali, and the remaining vaporized matter was observed by SEM. It consisted of continuous thin films.

Characteristics of Nylon Thin Films
As previously reported, the thickness of the thin films formed under the above conditions, determined by the SEM observation, were $100 \sim 200$ nm. Also, the examination results of the dissolution properties of the thin films shows that there existed two parts of the thin film which were soluble and insoluble in formic acid, a solvent for nylon. The analytical results of the thermal properties show that the soluble part had a melting point of 214℃ and a heat of fusion of 8 mcal/mg. These values are lower than the melting point of 225℃ and the heat of fusion of 17.4 mcal/mg for the original nylon. Based on the TGA results, it is presumed that the heat-reduction in weight of the formed thin films begun when the temperature of the films was considerably lower, that is, the films had such structures that are readily

decomposed at low temperature. In the FT-IR spectra, all of the characteristic absorption bands of nylon were observed, though the absorption intensities were rather different from those of the original nylon 6. In contrast to the nylon 6 specific gravity of 1.1519, that of the soluble part of the thin films was 1.3857, and that of the insoluble part was 1.2583. The X ray analysis reveals that the thin films are unoriented and noncrystalline. Based on these analytical results, it is presumed that nylon 6 is different from its original configuration and is a cross linked, uncross linked nylon 6 having a low molecular weight nearly the same as the original nylon 6. Moreover, a similar examination of nylon 66 was made. The results show that these nylons gave thin films similar to those formed in the case of nylon 6.

Types of Nylon and Adhesive Properties

The surfaces of PPODPTA woven fabrics were ion-plated with nylon 6 and 66 using the previously described conditions, and were RFL treated. The adhesive properties were then evaluated (shown in Fig.4) The nylon formed on the surfaces improved their adhesive properties. On the whole, the bonding strength and rubber coverage on the treated PPODPTA woven fabrics after peeling tests were relatively high.

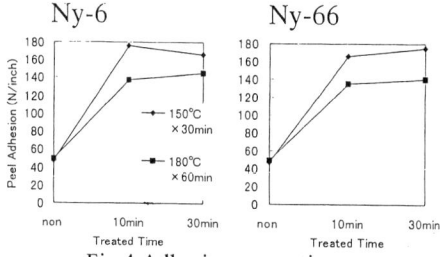

Fig.4 Adhesive properties

Comparison of Adhesive Properties due to Ordinary- and Over-Vulcanization

For, RFL treated nylon 6 and 66 fibers to be used in tire cords, the reduction of the adhesive properties, caused by over-vulcanization, was small and comparable to those due to the ordinary vulcanization. In the case of the aramid woven fabrics coated with the nylon thin films, the adhesive properties obtained by the over-vulcanization were relatively high, though a reduction was observed in some cases. The rubber coverage was also high.

Comparison of PPODPTA and PPTA

It has been also confirmed that PPTA takes almost the same behaviors as PPODPTA but with a difference in the adhesion level (shown in Fig.5).

More particularly, in the case of not only the low temperature plasma treatment but also the RFIP of the nylons, the adhesive properties of PPODPTA fibers were higher than those of PPTA fibers. As for the probable reason of this behavior, the affinities of nylon thin films to the RFL adhesive are estimated to be equal, and it is assumed that the adhesion properties of the nylon thin films to the PPODPTA substrate are

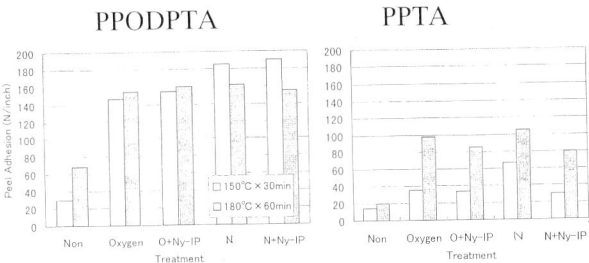

PPODPTA   PPTA

Fig.5 Comparison of PPODPTA and PPTA on adhesion level

higher than those to PPTA. In fact, PPODPTA exhibited very high rubber coverage, showing its cohesion failure. The PPTA showed very low rubber coverage, mainly causing failure at its interface. It may be concluded that the PPTA has relatively low-level adhesive properties though some improvement is seen when compared with the untreated PPTA.

Excimer Laser Irradiation (Ablation) Method

Changes in Fibers on Surface

Fig.6 shows the scanning electron micrographs of the irradiated aramid fiber surfaces after the laser pulse irradiation at 100 mJ/cm^2 with (a) non-irradiation, and (b) 10 shots. Small ripples were observed by the radiation. However, we found in this experiment that the ripples on the surface of PPODPTA are clearer than that of PPTA at the same conditions.

PPODPTA   **PPTA**

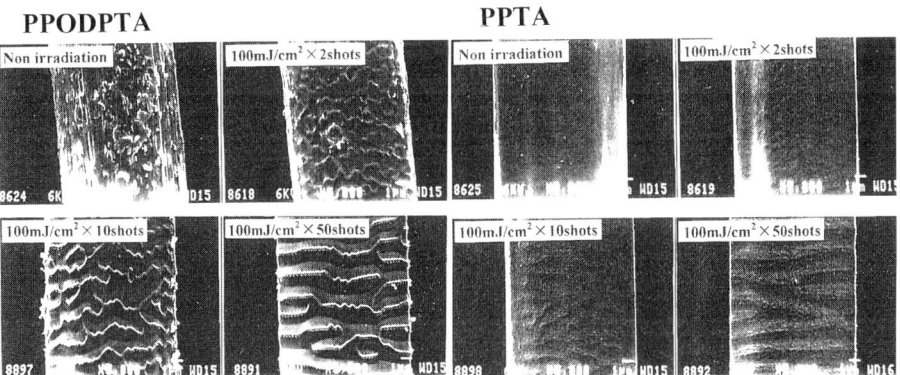

Fig.6   SEM micrographs of the fiber surfaces

Surface Chemical Composition studied by XPS Analysis

The content of O and C atoms at the aramid fiber surface after excimer laser irradiation was analyzed by XPS (ESCA). The ratio of O to C atoms decrease for PPODPTA and not change for PPTA. It may depend on the existing of ether bond of PPODPTA.

Adhesive Properties

Generally, as adhesives for fiber reinforced rubber composite, a first treating agent containing an epoxy compound having an affinity to the fibers and a second treating agent containing RFL are applied. The RFL comprises resorcinol-formaldehyde•-rubber latex and is reactive with the matrix rubbers through its methylol groups. Both of the treating agents, after they have been applied to the fibers, are cured with heat. The heating conditions exhibit a great influence on the adhesion performances of the fibers.

The authors have examined the application of an epoxy acrylate as the first treating agent. The acrylate compound, which is sensitive to UV rays, has a structure relatively similar to the epoxy compound used as the first treating agent. In order to reveal the effects of an excimer laser, the irradiation timing was examined. Before or after addition of a glycerol triglycidyl ether or its acrylate adducted-compound to the fibers, these were irradiated with an excimer laser (in the wet state) to compare the performances. In any case of the fibers, the KrF laser irradiation frequency was set that it slightly exceeded the threshold value and fine concave and convexes could be formed at the surfaces of the fibers. The irradiation was conducted at 100 mJ/cm^2×10 shots for the aramid fibers.

In the case that the laser irradiation was conducted beforehand, and then the first treating agent was applied and heated to be cured, differences between the adhesive properties of the fibers were observed. On the other hand, in the case that the compounds were applied and then cured by the laser irradiation, no changes in the adhesive properties were observed for the epoxy compound, but significant improvements in adhesive properties of the PPODPTA fibers were exhibited when the acrylate compound was used. However, in this case, no effects on the PPTA fibers were recognized. There was such tendency found with respect to the original and over-cured adhesive properties.

Irradiation conditions of KrF laser and adhesive properties

It was recognized that the adhesive properties can be enhanced when the epoxy acrylate is used as the first treating agent and cured with the KrF laser, followed by the RFL treatment and the heat curing. The adhesive properties were shown in Fig.7.

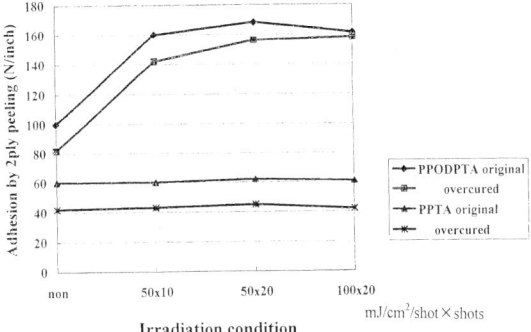

Fig.7 Adhesive properties of aramid fibers at various conditions

As to the observed rubber coverage at the peeled surfaces, PPODPTA presented the cohesive failure at about 50 mJ/cm^2 and also a desirable rubber coverage ratio. When the irradiation energy was increased, little changes in the apparent bonding strength of PPODPTA were shown, and the cohesive failure of the rubber was partially observed at the peeled surfaces. On the other hand, PPTA had a low bonding strength, showing the interfacial peeling and less rubber coverage.

Comparison of PPODPTA and PPTA

The aramid fibers PPODPTA and PPTA presented different adhesive properties. A reason was assumed as follows. In PPODPTA fibers, copolymer components compose crystalline domains independently of each other, and are oriented sufficiently. The copolymer components constitute crystalline domains, leaving almost no non-crystalline state in which molecules remain relaxed[5], whereas PPTA fibers comprise highly oriented skin layers and a core layer stacking up fibril groups in blocks. So there is a non-crystalline phase containing relatively many end groups of molecules between the blocks, which is likely to become fibril[6]. Such a difference in orientation of the non-crystalline phase seems to result in differences in susceptibility to ablation and in affinity of an epoxy acrylate compound to fibers, which is considered to be finally expressed as a difference in adhesion performance.

In addition, it is possible that the differences of the peeling-adhesion properties were caused by the different bonding energies of the fibers. In the PPODPTA fibers, the aromatic rings are linked together through amide and ether linkages; in the PPTA fibers, the aromatic rings are linked together only through amide linkages. The bonding energies are different and the fibers show different susceptibilities to the ablation. This might result in the different peeling-adhesion properties of the fibers.

When the laser irradiation was conducted after the epoxy acrylate was applied, a larger amount of fibrils were observed in the PPTA fibers at the peeled surface than in the PPODPTA fibers. It is estimated that the PPTA fibers present the low peeling-adhesion properties because they are readily fibrillated.

Conclusion

It has been revealed that the adhesive properties of the aramid fibers can be enhanced by low temperature plasma treating, and immediately after the treatment, carrying out the RFL treatment. It is important that the aramid fibers are subjected to the succeeding process as soon as possible before changes in the surface properties occur.

Based on the acknowledgment that nylon 6, one of the typical condensation type high polymers, can be formed into thin films by the RFIP method. The mechanism by low temperature plasma treatment on aramid fiber surface is shown in Fig.8.

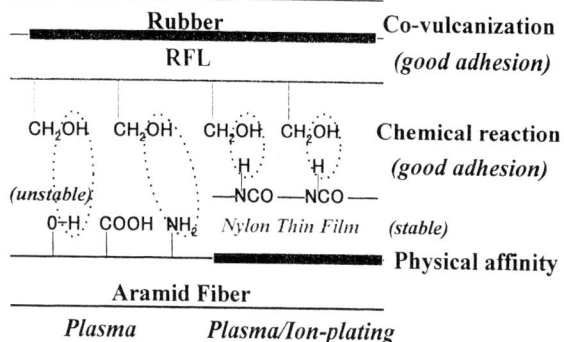

Fig.8 The mechanism by low temperature plasma treatment on aramid fiber surface

1) It has been confirmed that, as in the case of nylon 6, thin films made of un-oriented, noncrystalline nylons cross linked or uncross linked, having structures different from those of the original nylons and low molecular weight, are formed. According to the RFIP method, nylon thin films were formed on the fibers of which the surfaces are inactive, followed by treatment with RFL generally used fort rubber, thus having a high affinity to the nylon.

2) PPODPTA exhibited high adhesive properties such as, rubber coverage and bonding strength. As for PPTA, the bonding strength, though it was higher as compared with that of the untreated PPTA, was on a relatively low level. This is probably because the adhesion properties of PPODPTA and PPTA toward the nylon thin films were different. It is estimated that the RFIP method has a higher stability than the low temperature plasma treatment.

An examination was also made on the improvement of adhesion performances by a surface modification with an excimer laser. The following has been revealed. The mechanism by laser ablation on aramid fiber surface is shown in Fig.9.

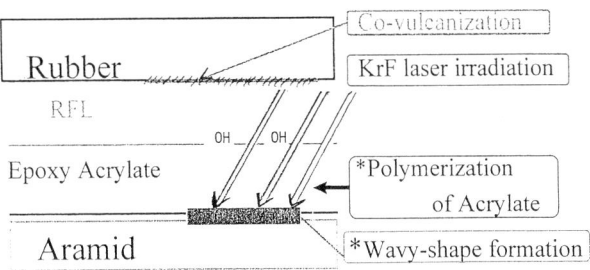

Fig.9 The mechanism by laser ablation on aramid fiber surface

3) When an acrylic acid adducted-compound (acrylate) is employed as first treating agent, and is cured by KrF laser irradiation, followed by a RFL treatment and heat-curing, the adhesive properties of the PPODPTA fiber are improved, but no improvements are observed for the PPTA fiber.

4) For the PPODPTA fibers, the copolymer components constitute crystalline domains, independently, which are sufficiently oriented substantially without amorphous regions having the molecular chains relaxed. Therefore, it is speculated that the high adhesive properties of the fibers could be obtained because the fibers were readily ablated, so that the applied acrylate compound enhanced the affinity of the fibers.

5) It is speculated that the relatively low adhesive properties of the PPTA fibers are attributed to a skin layer and a core layer composed of fibrils stacked into blocks. Amorphous phases containing a relatively large number of molecular chain ends are present between the blocks so that the fibers are readily fibrillated and are difficult to be ablated. Accordingly, the affinity of the fibers to the acrylate compound is insufficient.

We hope that the obtained knowledge will serve for the development of novel surface modification techniques for aramid fibers of which the surfaces are relatively inert.

References

1. Y. Ikada: "Materials Synthesis and Treatment by Plasma Reactions" 1984, p-177.

2. M. Furukawa, J. G. Dillard: ACS Meeting in Atlanta(1988.11.11)

3. E. Lawton: J. Appl. Polym. Sci. 18, 1557(1974)

4. T. Tomohisa, S. Tasaka, S. Seizo: Polymer Preprints, Japan 33, no 9, 2615(1984)

5. K. Thshiro, Y. Nakata, T. Izawa, M. Kobayashi, Y. Chatani, H. Tadokoro:
   Sen'i Gakkaishi 43 (1987) 627

6. R. J. Morgan: J. Polym. Sci., Polym. Phys. Ed. 21 (1983) 1757

*Macromol. Symp. **159**, 143–150 (2000)*

Instabilities in Thin Polymer Films: From Pattern Formation to Rupture

John R. Dutcher*, Kari Dalnoki-Veress[†], Bernie G. Nickel and Connie B. Roth

Department of Physics, University of Guelph, Guelph, Ontario, Canada N1G 2W1
[†]Present Address: Department of Physics, University of Sheffield, Sheffield, UK
S3 7RH

SUMMARY: Thermal fluctuations of the surfaces of thin polymer films can be amplified by the long-range van der Waals or dispersion force which acts across the film. When freely-standing polymer films are heated, this instability leads to the formation of holes. We have measured the formation and growth of holes in very thin, freely-standing polystyrene (PS) films to learn about the mobility of the confined polymer molecules. We have also symmetrically capped freely-standing PS films with thin, solid layers to probe the effects of mechanical confinement. Aggressive annealing of the trilayer films produces a novel in-plane morphology which can be understood in terms of the balance between the decrease in free energy associated with the dispersion interaction and the increase in free energy associated with the bending of the capping layers. The general nature of the morphology, and its reversibility, is demonstrated.

Introduction

Thin polymer films are being used increasingly in a wide variety of technological applications ranging from barrier layers in microelectronics to protective coatings to adhesives. In the continual push toward miniaturization of devices and minimization of material costs, the thickness of the films used in these applications has been, and will continue to be, reduced. As the films are made very thin, it is important to understand the physical properties of the polymer molecules confined to the films and, in particular, to determine if they differ from those in bulk. A particularly important physical property is the stability of the thin films. Soft materials such as polymers are characterized by bonding that is weak compared to conventional engineering materials such as metals and semiconductors. Because of this, large changes in the materials can be produced by relatively weak external stimuli such as the application of an external field or a change in temperature. In fact, modest temperature changes near room temperature can produce dramatic changes in the mechanical properties and the self-assembly of polymers. To be useful in technological applications, polymer films should have a morphology that either remains stable within an acceptable range of

          CCC 1022-1360/00/$ 17.50+.50/0

temperatures, or undergoes a transformation in morphology that is desirable, controllable and predictable. Reversible changes in morphology allow the possibility of device applications.

In the present manuscript, we will be concerned with thermally induced instabilities of thin polymer films. The most common type of thermally induced instability is the dewetting of a homopolymer film on an underlying substrate [1] or the formation and growth of holes in freely-standing or unsupported films. In both cases, holes form in the film due to two different mechanisms: amplification of thermal fluctuations of the free surface of the film by the attractive van der Waals or dispersion interaction across the film, resulting in *spontaneous* hole formation, and *nucleation* of holes at defects within the film.

The first measurements of hole growth in freely-standing polymer films were performed by Debrégeas *et al.* [2], in which they purposely nucleated holes in thick (thickness $h > 5$ μm) freely-standing polydimethylsiloxane films. Because the viscosity of the films was large ($\eta \sim 10^5$ P) compared with that of simple liquids such as water at room temperature ($\eta \sim 10^{-2}$ P), there was a qualitative difference in the growth of the hole radius: they observed exponential growth of the hole radius, instead of the linear growth of the hole radius which is observed for less viscous systems, e.g. soap films. In addition, they obtained direct evidence for uniform thickening of the film during hole growth, instead of the formation of a rim which is observed for soap films.

Below we describe two different but related studies: the thermal stability of freely-standing polystyrene (PS) films and freely-standing PS films that are mechanically confined by thin capping layers to form freely-standing trilayer films. Freely-standing PS films are of interest because very large reductions in the glass transition temperature T_g have been observed as the film thickness h is reduced to dimensions comparable to the root-mean-square end-to-end distance of the molecules [3]. For both of the studies described below, the driving force for the instability is the dispersion interaction which acts across the film and leads to spontaneous hole formation and growth in freely-standing PS films, and self-assembly and pattern formation in freely-standing trilayer films. The dispersion interaction between molecules is due to the Coulomb interaction between electric dipoles induced in molecules by the instantaneous electric dipole moment of neighbouring molecules [4]. The time average of the dispersion interaction is nonzero, even for molecules with no net electric dipole moment, and attractive. The dispersion interaction energy between isolated atoms and molecules separated by a distance R is described by a characteristic $U \sim R^{-6}$ dependence and is very weak at distances that are considerably greater than the equilibrium atomic spacing in solids. However, for collections of molecules, for which the dispersion interaction between all molecules must be accounted, the net dispersion interaction can be considerable over much larger distances. For example, consider a thin film of thickness h surrounded by vacuum. In this case [5], in which there is symmetry about the midplane of the film, the dispersion interaction across the film is always attractive and the strength of this interaction energy has a characteristic dependence on the film thickness h given by $U \sim h^{-2}$. Therefore, the dispersion interaction across a thin film is appreciable over distances which are large compared with the equilibrium atomic separation of solids. Typically, it is necessary to account for the dispersion interaction across a thin film for film thicknesses of the order of 100 nm. A rigorous treatment requires the use of the retarded form of the interaction potential for film thicknesses greater than 10 nm [4].

Since the driving force for the instabilities probed in the present study is the dispersion interaction, the present measurements are necessarily limited to very thin polymer films. However, it is possible to think more generally about the driving force for the instabilities, e.g. the application of an external field, which can be appreciable over much larger distances than the dispersion force. The observations obtained in the present study can then be extended to much thicker films.

Sample Preparation

For explicit details of the preparation of samples used in the studies described below, the reader is referred to the original papers [6,7]. We wish only to mention the specific steps in the preparation of freely-standing polymer films that we regard as being very important to achieve reproducible results. The basic scheme is to dissolve high molecular weight, narrow distribution polymer in a good solvent and to spincoat the solution onto clean glass slides or mica substrates. We are very careful to maintain the same post-deposition thermal treatment for all samples to ensure reproducibility of the measured results. The films on the substrates are annealed at a temperature well above the bulk glass transition temperature T_g for an extended period of time to drive off residual solvent and to allow the polymer chains to relax. The films on the substrates are then cooled to room temperature at a constant, slow rate, creating films with a well-defined thermal history. The films are then floated onto a distilled water surface and captured across a 4-mm diameter hole on a stainless steel sample holder, creating the freely-standing polymer films.

Freely-standing trilayer films composed of a variety of different materials were prepared to investigate the effects of mechanical confinement on self-assembly of polymer films. Either SiO_x or Au was evaporated onto freely-standing polymer films. In addition, all-polymer trilayer films were created by spincoating a bilayer sample onto clean glass slides and using the water transfer technique to deposit a third polymer layer and create the freely-standing films.

Hole Formation and Growth in Freely-Standing Polystyrene Films

We have used two experiments, optical microscopy and a differential pressure experiment, to study the formation and growth of holes in freely-standing polystyrene films.

In the optical microscopy study [6], we studied holes that formed spontaneously upon heating the films and holes that were purposely nucleated using a heated scanning tunneling microscope tungsten tip coated with a layer of SiO_x. By placing the films into the hot stage of a reflected light optical imaging system and heating the films to $T = 115$ °C (a temperature well above T_g), we have measured the radius of a hole in each film as a function of time.

In Fig. 1, we show a plot of the natural logarithm of the hole radius R, normalized to the radius R_0 first measured for each hole, versus time t for six representative films ranging in film thickness from $h = 96$ nm to $h = 372$ nm. As in the study of Debrégeas et al. [2], the plot of $\ln(R/R_0)$ versus t is linear for each film, corresponding to exponential growth of the hole radius: $R(t) = R_0 e^{t/\tau}$, where τ is the characteristic growth time of the hole.

In Fig. 2, we show a plot of the logarithm of τ versus film thickness h for all of the 25 films used in the hole growth study. For a given film thickness, we found that the τ value was reproducible to within ± 15%. The most striking feature of this data is the large decrease in τ with decreasing h: τ decreases by a factor of 33 as h is reduced by a factor of 3.9. The polymer viscosities can be inferred from the τ values and were of the order of 10^{10} P which is 5 orders of magnitude larger than in the studies of Debrégeas et al. [2]. We have shown that this experimental result can be understood in terms of the reduction in polymer viscosity with increasing shear strain rate (a nonlinear viscoelastic phenomenon known as shear thinning) [6].

To probe the onset of hole formation in very thin freely-standing PS films, we have developed a differential pressure experiment [8]. In this experiment, we have a pressure cell which is divided into two compartments with a small hole connecting the two compartments. The

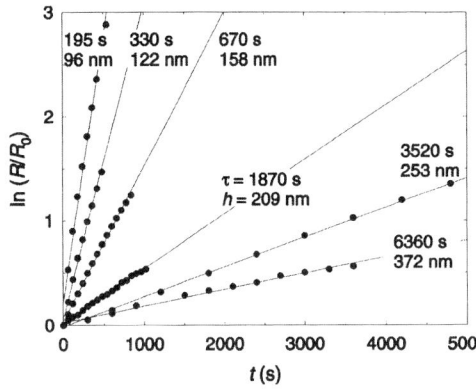

Fig. 1: Natural logarithm of the hole radius R, normalized to the value of the radius R_0 first measured for each hole, as a function of time for six representative PS films ranging in film thickness from $h = 96$ nm to 372 nm [6].

freely-standing PS film is placed across the hole, forming a barrier between the two compartments. A small (10^{-4} atm) constant pressure difference is applied and maintained across the freely-standing film using a stepper-motor driven piston attached to one compartment of the pressure cell. As the film is heated, holes form in the film due to the thermal instability. This causes gas to flow across the film, which is detected as a monotonic drift in the piston position as the system attempts to maintain the constant pressure difference across the film. The experiment was performed using films with thickness $h = 65$ nm, which have a glass transition temperature T_g which is reduced from the bulk value by 40 °C [3]. For these films, we find that hole formation occurs only at temperatures comparable to or slightly less than the bulk value of T_g. Therefore, for the freely-standing PS films, the glass transition temperature occurs at a much lower temperature than the motion of entire polymer chains, as measured by hole formation.

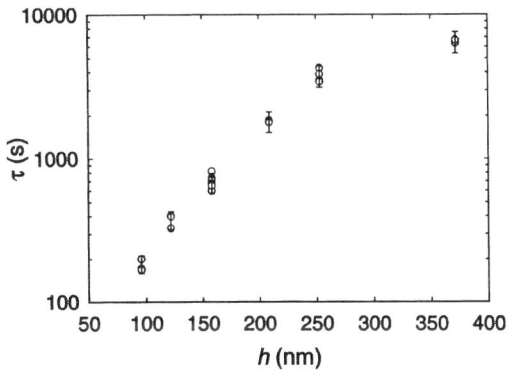

Fig.2: Characteristic growth time τ versus film thickness h for freely-standing PS films [6].

Dispersion-Driven Morphology in Freely-Standing Trilayer Films

A simple strategy for improving the thermal stability of freely-standing polymer films is to add thin, solid capping layers to both surfaces of the freely-standing films, creating freely-standing trilayer films [7]. As expected, hole formation in the freely-standing trilayer films is suppressed when the films are heated to temperatures at which hole formation occurs in the freely-standing polymer films. However, by merely heating the freely-standing trilayer films to considerably higher temperatures, the system self-assembles to create a novel in-plane morphology, consisting of long, parallel domains with a well-defined periodicity. The in-plane morphology obtained upon aggressive annealing is shown in Fig. 3 for a variety of freely-standing trilayer films, as well as a variety of polymer films supported on substrates and capped with a thin, solid layer. The general nature of the phenomenon can be seen by the observation of the same type of in-plane morphology in a wide variety of different freely-standing and supported film systems. The formation of the long parallel domains can be understood in terms of a competition between the attractive dispersion interaction which would like to bring the outer surfaces of the film together and the energy cost associated with elastically bending the capping layers. Parallel domains are obtained because it requires less energy to deform a sheet (each capping layer) in parallel bends than in uncorrelated deformations. We have developed a simple model based on linear stability analysis that allows us to correctly predict the dependence of the periodicity λ of the morphology on the individual layer thicknesses: $\lambda \sim L^{3/4} (h + 2L)$, where h is the polymer layer thickness and L is the capping layer thickness. The success of the model prediction can be seen in Fig. 4, in which $\lambda L^{-3/4}$ is plotted versus $(h + 2L)$ for 16 different $SiO_x/PS/SiO_x$ freely-standing trilayer films with different values of h and L. All of the data lie along a straight line, as predicted by the model.

We have also achieved reversibility of the morphology. Specifically, for the PS/PI/PS system, the in-plane morphology was created by annealing PS/PI/PS freely-standing trilayer films and then removed by changing the dispersion force acting across the film by bringing the freely-standing film into contact with Si [7].

It is worth noting that the present study of self-assembly and pattern formation in homopolymer films is a simplified version of a previous study of the thermal stability of thin films composed of two immiscible polymers (PS and PMMA) supported on Si and capped with a thin solid layer of SiO_x [9]. The polymers were dissolved in a common solvent and

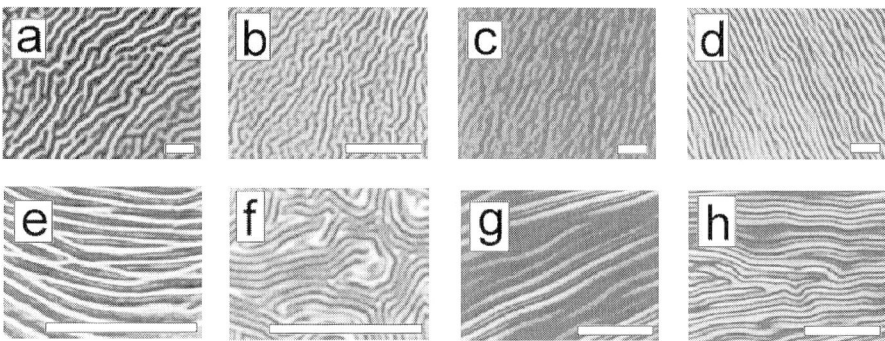

Fig. 3: Optical microscope images obtained for annealed freely-standing trilayer films (a – d) and annealed supported films with capping layers (e – h). In particular, the samples are (a) $SiO_x/PS/SiO_x$, h = 121 nm, L = 22 nm; (b) $SiO_x/PMMA/SiO_x$, h = 120 nm, L = 40 nm; (c) Au/PS/Au, h = 70 nm, L = 15 nm; (d) PS/PI/PS, h = 50 nm, L = 70 nm; (e) Au/PS/SiO$_2$, h = 40 nm, L = 30 nm; (f) $SiO_x/PS/Si$-H, h = 70 nm, L = 40 nm; (g) PS/PI/Si-H, h = 50 nm, L = 70 nm; (h) PS/PI/SiO$_x$/Si, h = 50 nm, L = 70 nm, where PS is polystyrene with molecular weight M_w = 767 × 10^3, PMMA is poly (methyl methacrylate) with M_w = 1.22 × 10^6, PI is polyisoprene with M_w = 414 × 10^3, and Si-H is hydrogen-terminated Si. For each image, a horizontal bar corresponding to 20 μm is indicated.

deposited using spincoating. The molecular weights of the polymers were sufficiently small and the solvent was sufficiently volatile such that appreciable phase separation did not take place during the deposition of the films, i.e. the blend films were homogeneous on all length scales that can be accessed using optical microscopy. The films were then capped with a thin layer of SiO_x and then heated to temperatures greater than the glass transition temperatures of

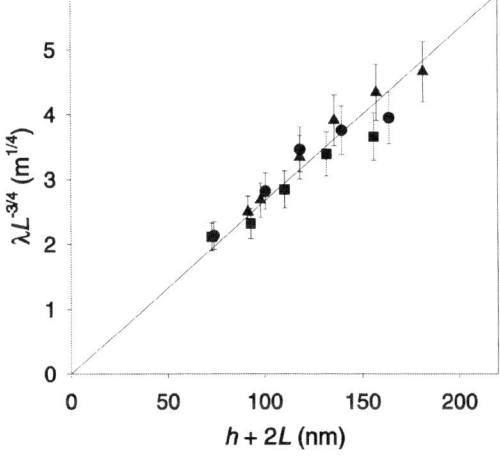

Fig. 4: Plot of $\lambda L^{-3/4}$ versus trilayer film thickness (h + $2L$) for 3 series of $SiO_x/PS/SiO_x$ freely-standing trilayer films with capping layer thicknesses L = 17.7 nm (■), 21.6 nm (●) and 30.4 nm (▲) [7].

the two polymers, allowing phase separation to take place. The driving force for the phase separation process was the immiscibility of the two polymers. It was found that the resulting phase separation had two distinct morphologies. For thick capping layers, the phase separation was lamellar, since the PMMA preferred the SiO_x surfaces, with no in-plane structure within the film. For sufficiently thin capping layers, an in-plane morphology similar to that shown in Fig. 3 was obtained. The transition between the two morphologies was explained in terms of a balance between the change in free energy associated with forming interfaces between the PS-rich and PMMA-rich domains and the change in free energy associated with the elastic bending of the capping layer.

Conclusion

We have described the results of two different studies related to the thermal stability of thin, freely-standing polymer films. These studies have revealed new information regarding the mobility of polymer molecules confined to thin films, as well as the observation and understanding of a novel, in-plane morphology of a general nature.

References

1. C. Redon, F. Brochard-Wyart and F. Rondelez, Phys. Rev. Lett. **66**, 715 (1991).
2. G. Debrégeas, P. Martin and F. Brochard-Wyart, Phys. Rev. Lett. **75**, 3886 (1995).
3. J.A. Forrest, K. Dalnoki-Veress, J.R. Stevens and J.R. Dutcher, Phys. Rev. Lett. **77**, 2002 (1996); J.A. Forrest, K. Dalnoki-Veress and J.R. Dutcher, Phys. Rev. E **56**, 5705 (1997).
4. J. Mahanty and B.W. Ninham, *Dispersion Forces*, Academic Press, London, 1976.
5. J. Israelachvili, *Intermolecular & Surface Forces*, Second Edition, Academic Press, London, 1992.
6. K. Dalnoki-Veress, B.G. Nickel, C. Roth and J.R. Dutcher, Phys. Rev. E **59**, 2153 (1999).
7. K. Dalnoki-Veress, B.G. Nickel and J.R. Dutcher, Phys. Rev. Lett. **82**, 1486 (1999).
8. C.B. Roth, K. Dalnoki-Veress, B.G. Nickel and J.R. Dutcher, unpublished.
9. K. Dalnoki-Veress, J.A. Forrest and J.R. Dutcher, Phys. Rev. E **57**, 5811 (1998).

Surface Modification of Polymeric Materials by Remote Plasma

Norihiro Inagaki

Dept. of Materials Science, Faculty of Engineering, Shizuoka University

3-5-1 Johoku, Hamamatsu, 432-8561 Japan

Phone & Fax. : 81-53-478-1161, E mail. tcninag@mat.eng.shizuoka.ac.jp

SUMMARY: Remote plasma treatment has been proposed here as a special technique to modify polymer surfaces.   A concept for the remote plasma treatment rises from the anticipation that radicals rather than electrons and ions in plasma contribute to introduction reactions of functional groups and the radicals have longer lifetime than electrons and ions.   As a result, effective surface modification at a position far from the plasma zone (remote plasma treatment) will be done because radicals rather than electrons and ions become a predominant species for the modification.

Fluoropolymers,   poly(tetrafluoroethylene),   tetrafluoroethylene-hexafluoropropylne copolymer,   and tetrafluroethylene-perfluoroalkylvinylether copolymer, were modified by the remote hydrogen plasma.   Their surface morphology and surface chemistry were demonstrated.   An application of the remote hydrogen plasma treatment for improvement of adhesion between copper metal and the fluoropolymers was also demonstrated.

Introduction

Plasma treatment is a technique to modify surface properties and surface topology of polymeric materials.   Modification of polymeric surfaces from hydrophobic to hydrophilic or from hydrophilic to hydrophobic is easily accomplished by means of exposing their surfaces to plasmas for a few minutes.   Table 1 shows typical results of hydrophilic and hydrophobic modification of polyethylene surfaces by eight different plasmas[1].   In these modification processes, hydrophilic or hydrophobic groups are introduced on the polyethylene surfaces by reactions with plasma (introduction of functional groups).   Radicals in the plasmas play an important role in the introduction reactions.   The radicals abstract hydrogen atoms from polymer chains to form carbon radicals in the polymer chains, and the carbon radicals react with other radicals in plasma to form hydrophilic or hydrophobic groups on the polymer surface.   These are essential reactions for the introduction reactions of functional groups.   On the other hand, the other species besides in the plasmas, electrons and ions, bombard the polymer surface during the process of introduction reactions, and initiate degradation reactions on the polymer surface.   As a result, degradation products with low-molecular weight and an injur-

152

ed layer (weak boundary layer) are formed on the polymer surface.

    Therefore, in the surface modification by plasma, two different processes, introduction of functional groups and degradation, occur simultaneously on the polymer surfaces. In the sense of chemistry, the degradation of polymer surfaces is an uncalled-for process in the surface modification. As long as plasma is used as reactive species for the modification, degradation on the polymer surfaces is never avoided from the surface modification by plasma treatment. Is there any amelioration for plasma treatment without degradation of polymer surfaces? We propose here remote plasma as a possible amelioration.

Table 1 Water contact angle on polyethylene surfaces treated with plasmas

Kind of plasma	Water contact anlge (degree)
-	99
CO	16
CO_2	8
NO	25
NO_2	37
O_2	35
SF_6	117
CF_4	121
C_2F_6	114

Introduction of functional groups and degradation on polymer surface

Table 2 compares effects of plasma treatment on polyethylene (PE) and polypropylene (PP). PE and PP surfaces were modified with Ar and O_2 plasmas and then adhered with epoxy adhesive[2]. Peel strength for the PE/epoxy adhesive and PP/epoxy adhesive systems was evaluated from the viewpoint of effects by the plasma treatment. Surface modification by the Ar and O_2 plasma was effective for PE in improving the adhesion with epoxy adhesive. On the other hand, the Ar and O_2 plasmas were no effective for PP in the adhesion improvement. This disparity between PE and PP in effects of the plasma treatment

Table 2 Effects of plasma treatment on adhesion between polymers and epoxy adhesive

Polymers	Kind of plasma	90° Peel strength (N/25 mm)
PE	-	0.1
	Ar	17.4
	O_2	28.6
PP	-	0
	Ar	0.1
	O_2	0.1

may be due to not only introduction reactions of functional groups but also degradation reactions. When PE and PP were exposed to Ar plasma, weight loss and degradation products were formed on their surfaces.

Table 3 shows weight loss and water-soluble degradation products formed on PE and PP surfaces, when their surfaces were exposed to Ar plasma at 25W for 2 min[2]. Weight loss and water-soluble product formed on the PE surface were very small. On the other hand, weight loss and degradation products on the PP surface were about ten times larger than those from PE. Therefore, the PP surface is easy to be degrad-

Table 3 Degradation products formed by Ar plasma treatment (at 25W for 2 min)

Polymers	Weight losss by Ar plasma treatment ($\mu g/cm^2$)	Fromation of water-soluble products ($\mu g/cm^2$)
PE	2	1
PP	16	11

ed by the Ar plasma, and as a result, the PP surface is deposited by degradation products. The deposition of degradation products on the PP surface may be a factor to disturb the adhesion with epoxy adhesives.

The two tables (Tables 2 and 3) indicate an important aspect on modification of polymer surfaces by plasma. Actions by plasma on polymer surfaces contain an introduction reaction of functional groups and a degradation reaction. The introduction reaction does not isolate from the degradation reaction but coexists always with the degradation reaction during the surface modification process (Fig. 1). How much the introduction reaction or the degradation reaction occurs depends on the nature of polymers as well as plasma conditions. Even when

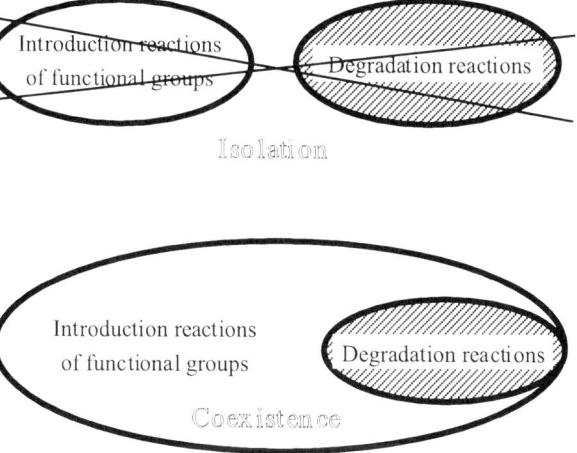

Fig.1: Aspect of plasma treatment process

the same plasma is used for the surface modification, different effects of the plasma treatment will appear in the case of polymers having different plasma susceptibility, as an example, PE and PP. We can not get out of the degradation process, as long as plasma is used as a reactive species for the surface modification. In the sense of chemistry, the degradation reaction is uncalled-for one in the surface modification. How to minimize the degradation reaction is an important factor to make a success of effective surface modification.

Surface modification by remote plasma

We propose here a special technique, "Remote Plasma Treatment" to minimize degradation reactions occurring during the plasma treatment. A concept for the remote plasma treatment rises from the anticipation that radicals rather than electrons and ions in plasma contribute to introduction reactions of functional groups and the radicals have longer lifetime than electrons and ions[3].

Plasma contains electrons, ions, and radicals that can interact with polymer surfaces. The interaction leads to introduction reactions of functional groups on the surfaces and degradation reactions of the polymer surfaces. The introduction reactions are mainly initiated by interaction with radicals. On the other hand, the degradation reactions are initiated by the bombardment of electrons and ions. If radicals alone are separated from the plasma and then, interact with the polymer surface, the introduction reactions will occur on the polymer surfaces without degradation reactions. Fig. 2 shows typical elemental reactions in hydrogen plasma[4-8]. There is a small difference in rate constant between the generation reactions of hydrogen radical (H) and hydro-

1. Generation of active species

$$e + H_2 \xrightarrow{k_1} H_2^+ \qquad k_1 = 2.32 \times 10^{-11} cm^3/s$$

$$e + H_2 \xrightarrow{k_2} 2H \qquad k_2 = 4.49 \times 10^{-12}$$

2. Disappearance of active species

$$H_2^+ + e \xrightarrow{k_3} H_2 \qquad k_3 = 5.66 \times 10^{-8}$$

$$2H + H_2 \xrightarrow{k_4} 2H_2 \qquad k_4 = 8.3 \times 10^{-33}$$

Fig. 2: Typical elemental reactions in hydrogen plasma

gen ion (H_2^+), but there is a large difference between the disappearance reactions of the hydrogen radical and hydrogen ion.   The rate constant (k_4) of the disappearance of hydrogen radicals is negligibly small compared with that of hydrogen ions (k_3).   As a result, the hydrogen radical will be a predominant component if the hydrogen plasma elapses after discharge.

Fig. 3 shows a special reactor to separate hydrogen radicals from hydrogen plasma[9).   The reactor consisted of a cylindrical Pyrex glass tube (45 mm diameter, 1000 mm long) and a columnar stainless steel chamber (300 mm diameter, 300 mm height).   The Pyrex glass tube had two gas inlets for the injection of hydrogen and argon gases and a copper coil for the energy input of rf power (13.56 MHz frequency) at an end of the tube.   The Pyrex glass tube at the other end was jointed with the stainless steel chamber in a manner of Viton O ring flange.   The stainless steel chamber was contained a vacuum system of a combination of a rotary pump and a diffusion pump.   Polymer samples for modification were positioned in the Pyrex glass tube at a constant distance of 800 mm from the center of the copper coil.

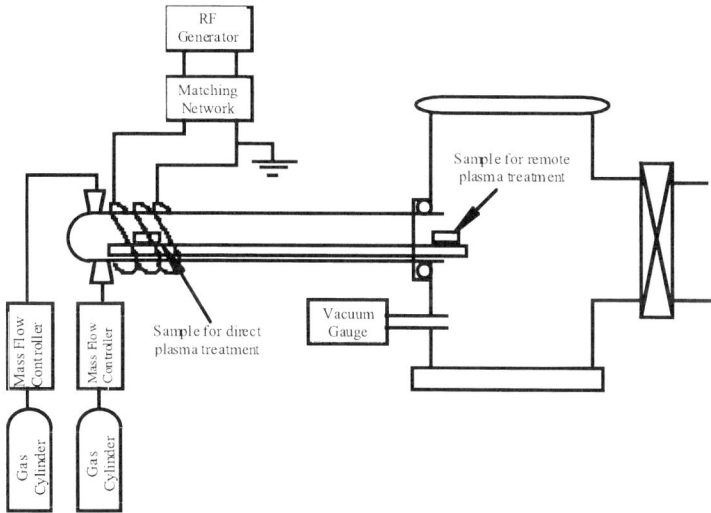

Fig. 3: Schematic presentation of reactor for remote plasma treatment

Hydrogen gas (10 cm^3/min) was injected from the gas inlet, and discharged at the copper coil to make hydrogen plasma.   Active species, electrons, hydrogen ions, and hydrogen radicals, in the hydrogen plasma traveled in the direction of the stainless steel chamber with hydrogen gas stream and streamed out from the reactor. At a distance of 800 mm from the plasma zone, hydrogen radicals will be a predominant component and electrons and hydrogen ions will be minor component, because of a large difference in lifetime. This is a basic concept of "Remote Plasma Treatment". Difference between the remote and conventional plasma treatments is the relative position of polymer samples from the plasma zone. We call here the conventional plasma treatment "direct plasma treatment".

Comparison of remote and direct plasma treatments

Fig. 4 shows SEM pictures of poly (tetrafluoroethylene), PTFE, surfaces treated with the remote and direct hydrogen plasmas (at 100W, at 13.56 MHz)[10]. The PTFE surface treated with the remote hydrogen plasma (The PTFE sample was set up at a distance of 80cm far from the hydrogen plasma zone.) showed a scabrous surface that contained many micropores of $0.2 - 0.4$ μm wide and $1 - 2$ μm long. This SEM picture is similar surface morphology to

the original PTFE. On the other hand, the PTFE surface treated with the direct hydrogen plasma (The PTFE sample was set up in the hydrogen plasma zone.) showed distinctly different one in the surface morphology from the remote hydrogen plasma-treated PTFE surface. The surface roughness was increased to $1 - 2$ μm wide and $3 - 7$ μm long. This comparison indicates that less degradation occurred in the remote hydrogen plasma, and heavy degradation reaction occurred in the direct hydrogen plasma.

Original PTFE

PTFE treated with remote hydrogen plasma at 100W for 120s

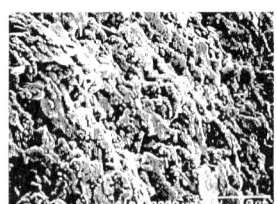

PTFE treated with direct hydrogen plasma at 100W for 30s

The two PTFE surfaces treated with the remote and direct hydrogen plasmas were analyzed with XPS (Table 4)[9]. The PTFE surface treat-

Fig. 4: SEM pictures of PTFE surfaces treated with remote and hydrogen plasmas

ed with the remote hydrogen plasma showed a large decrease in F/C atom ratio and increase in O/C atom ratio. Similarly, the PTFE surface treated with the direct hydrogen plasma showed similar decrease in F/C atom ratio and increase in O/C atom ratio.

Table 4   Atom composition of remote and direct hydrogen-plasma-treated PTFE surfaces

| Plasma | Atom composition on treated PTFE surfaces | |
	F/C atom ratio	O/C atom ratio
Remote H_2	0.41	0.12
Direct H_2	0.60	0.07
(Untreated)	1.90	0.05

This result indicates that defluorination and oxidation occurred in the remote hydrogen plasma treatment as well as in the direct hydrogen plasma.   Even when the PTFE surface did not touch directly to the hydrogen plasma, defluorination occurred effectively.   The formation of oxygen moieties on the PTFE surface by the remote hydrogen plasma may be due to post reactions of carbon radicals with oxygen in air after the plasma treatment.   Carbon radicals are formed by means of abstraction reactions of fluorine atoms from the PTFE surface by hydrogen atoms and by means of the bond scission of the C-F and C-C bonds by electron and ion bombardment.

XPS spectra gave us useful information on the chemical composition of the PTFE surfaces treated with the remote and direct hydrogen plasmas.   Fig. 5 shows typical XPS (C_{1s}) spectra for the two PTFE surfaces[9].   The C_{1s} spectra for the untreated PTFE showed a peak at 292.5 eV due to CF_2CF_2 groups.   The remote- and direct-hydrogen plasma-treated PFTE showed, as shown in Fig. 5, complex C_{1s} spectra having two peaks.   The spectra were decomposed into five components appearing at 285.9-296.0, 287.6-287.9, 289.6-289.8, 292.5, and 293.7-294.1 eV.   These five components are assigned to the $\underline{C}H_2CHF$ groups at 285.9-296.0 eV, $\underline{C}HFCH_2$ and C=O groups at 287.6-287.9 eV, $\underline{C}H(OR)CHF$ groups at 289.6-289.8 eV, $\underline{C}F_2CF_2$ groups at 292.5 eV, and $\underline{C}F_3CF_2$ and $\underline{C}F(OR)_2CF_2$ groups at 293.7-294.1 eV. The underlined carbon atoms are objective atoms for the assignment.   This analysis shows that there is no difference in chemical composition between the PTFE surfaces treated with the remote and direct hydrogen plasmas.   The remote hydrogen plasma as well as the direct hydrogen plasma has a high capability of the surface modification of PTFE.   Hydrogen substitution on the PTFE surface occurs in the modification process by the remote hydrogen plasma.

The O_{1s} spectra also gave us useful information on oxygen moieties formed on the PTFE surfaces. The spectra are not represented here for shortness' sake. The O_{1s} spectra were decomposed into two components, O=C and O-C groups, appearing at 532.3-532.4 and 533.6-533.7 eV, respectively. This analysis shows that oxidation occurred on the carbon atoms to form C=O and C-O groups.

From these results, we conclude that the remote hydrogen plasma leads to effective modification of PFTE surfaces without heavy degradation on their surfaces.

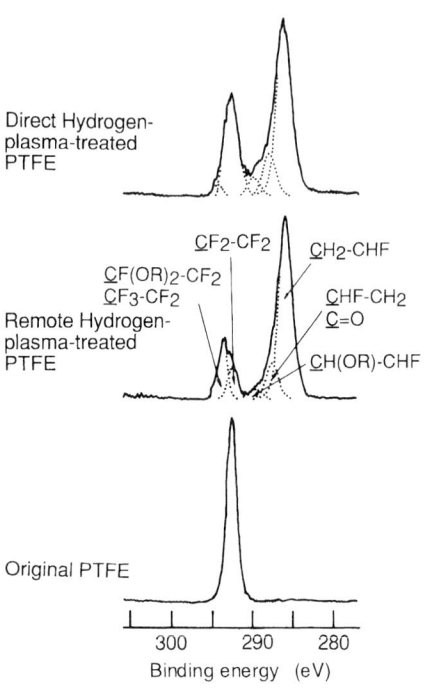

Fig. 5: C_{1s} spectra for PTFE surfaces treated with remote and direct hydrogen plasmas

Application of Remote Hydrogen Plasma Treatment

– Copper Metallization on Fluoropolymer Surfaces -

Fluoropolymers; poly(tetrafluoroethylene), PTFE; tetrafluroethylene-hexafluoropropylene copolymer, FEP; and tetrafluroethylene-perfluoroalkylvinylether copolymer, PFA; show distinguished physical and chemical properties such as high-temperature resistance and chemical resistance as well as hydrophobic properties. In addition to these distinguished properties, these polymers show excellent electrical-properties which are high resistivity of more than 10^{18} Ωcm, a low dielectric constant of 2.1, and a low dissipation factor of less than 2×10^{-4}. Furthermore, FEP and PFA can be fabricated easily in conventional melt processes because of amorphous structure. From this aspect, we believe that the fluoropolymers may be an outstanding insulator from electrical currents with high frequency (GHz), for example, a printed wiring board for integrated circuits and an insulator for electrical wires. To apply the fluoropolymers for the electrical materials, copper metallization on the fluoropolymer surfaces was investigated here using the surface modification by the remote hydrogen plasma.

The fluoropolymer (PTFE, FEP, and PFA) sheets (50 μm thickness) were treated with the remote hydrogen plasma (at 13.56 MHz and at 25 – 100W), and then metallized by a combination of an electroless- and electro-platings. The total thickness of the deposited copper metal on the fluoropolymer surfaces was about 30 μm (0.2 μm thickness by the electroless plating and 30 μm thickness by the electroplating). The adhesion between the copper metal layer and the fluropolymer sheet was evaluated as T-type peel strength at a peel rate of 10 mm/min.

Fig. 6 shows typical results for the peel strength as a function of the plasma exposure time in the remote hydrogen plasma treatment[11]. PFA sheets (5 mm wide) were used as a fluoropolymer substrate for the copper metallization experiment. The peel strength, as shown in Fig. 6, increased with increasing the plasma exposure time, reached a maximum of 101 mN/5mm, and then leveled off. This indicates that the surface modification by the remote hydrogen plasma treatment contributes improvement of adhesion between copper metal and PFA. The surface modification by the direct hydrogen plasma also was effective in improving the adhesion, but the improvement was not as

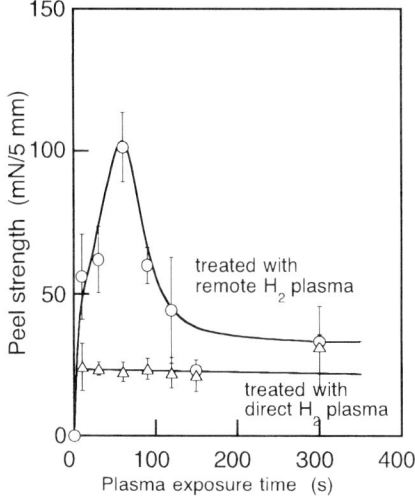

Fig. 6: Peel strength for copper metal/ PFA composites as a function of plasma exposure time in remote and direct hydrogen plasma treatment processes.

powerful as that by the remote hydrogen plasma (shown in Fig. 6). The peel strength was increased from 0 to 24 mN/5mm by the direct hydrogen plasma treatment, but further increase was never seen. This indicates that the direct hydrogen plasma was not effective in modifying the PFA surfaces, although defluorination occurred effectively on the PFA surfaces. The ineffectiveness may be due to heavy degradation reactions on the PFA surface.

XPS (C_{1s}, F_{1s}, and Cu_{2p}) spectra for surfaces of the PFA and copper metal layers failed from the copper metal/PFA system showed a mechanism of the failure mode from the composites[11]. The PFA and copper metal layers peeled off from the copper metal/PFA system showed similar C_{1s} and F_{1s} spectra and no Cu_{2p} spectrum. The C_{1s} spectra contained an intense and sharp peak at 292.8 eV and a small but widely distributed peak at 287 – 284 eV. The F_{1s} spectra contained a symmetrical peak at 690 eV. These spectra show that the PFA and copper metal layers peeled off from the copper metal/PFA system are similar in chemical composition. The failure occurred in an inner layer of plasma-treated PFA polymer rather than at the interface between the copper metal and PFA layers.

The other fluoropolymers, PTFE and FEP, also were modified by the remote hydrogen plasma, and the adhesion with copper metal was improved effectively (Table 5)[12]. The surface modification by the remote hydrogen plasma contributes to improvement of the adhesion between copper metal and fluoropolymers.

Table 5   Peel strength for copper metal/fluoropolymer systems

Fluoropolymers	Peel strength (mN/5mm)	
	Untreated	Treated with remote H_2 plasma
PTFE	7.5	92
FEP	0	195
PFA	0	101

Conclusion

We have proposed here a special technique, "Remote Plasma Treatment" to minimize degradation reactions occurring in the plasma treatment. A concept for the remote plasma treatment rises from greatly longer lifetime of radicals than electrons and ions. Capability of the remote hydrogen plasma has been demonstrated in modifying fluoropolymer surfaces and applying the modification into the copper metallization of the fluoropolymer surfaces. We believe that the remote hydrogen plasma treatment is a preferable technique for surface modification of fluoropolymers.

References

1. N. Inagaki, S. Tasaka, J. Ohkubo, *J. Appl. Polym. Sci., Appl. Polym. Symp.* **46**, 399 (1990)
2. N. Inagaki, in *Plasma Surface Modification and Plasma Polymerization*, Technomic Pub., Lancaster (Pennsylvania), 1996
3. E. E. Kunhardt, L. H. Luessen, Eds., in *Electrical Breakdown and Discharge in Gases*, Plenum, New York, 1983

4. M. J. Kushner, *J. App. Phys.* **63**, 74 (1988)
5. C. –H. Chou, T. –C. Wei, *J. Appl. Phys.* **72**, 870 (1992)
6. T. –C. Wei, T. –C. Phillips, *J. Appl. Phys.* **74**, 825 (1993)
7. H. Rau, F. Phichat, *J. Phys. D: Appl. Phys.* **26**, 1260 (1993)
8. J. Deson, F. Halous, C. Lalo, C. Rousseau, V. Veniard, *J. Phys. D: Appl. Phys.* **27**, 2320 (1994)
9. Y. Yamada, T. Yamada, S. Tasaka, N. Inagaki; *Macromolecules* **29**, 4331 (1996)
10. N. Inagaki, S. Tasaka, T. Umehara;,*J. Appl. Polym. Sci.* **71**, 2191 (1999)
11. N. Inagaki, S. Tasaka, K. Mochizuki, *Macromolecules* **32**, 8566 (1999)
12. N. Inagaki, S. Tasaka, Y. W. Park, *J. Adhesion Sci. Techonol.* **12**, 1105 (1998)

Predicting the Formation, Structure and Elastomeric Properties of End-Linked Polymer Networks

R.F.T. Stepto[*], J.I. Cail and D.J.R. Taylor

Polymer Science and Technology Group, Manchester Materials Science Centre, UMIST and University of Manchester, Grosvenor St., Manchester, M1 7HS, UK

SUMMARY: The molecular structures and macroscopic properties of network polymers depend more closely on reactant structures (molar masses, functionalities, chain flexibilities) and reaction conditions (dilution, proportions of different reactants) than do those of linear polymers. To understand and predict elastomeric properties, it is important to be able to model, statistically, the molecular growth leading to network formation. A new Monte-Carlo network polymerisation algorithm has been developed, using Flory-Stockmayer random-reaction statistics with intramolecular reaction allowed on a correctly weighted basis. The algorithm simulates, as a function of extent of reaction, the formation of *all of the connections* in a reaction mixture and counts *all* the ring structures. It also enables polymerisations and network structures to be simulated efficiently up to *complete* reaction. Comparisons of predictions from the algorithm with experimental data from end-linking polymerisations show the importance of accounting for the *whole* distribution of sizes of ring structure in determining reductions in elastic modulus. An important new factor, x, is introduced in the interpretation of experimental data. It is the fractional loss in elasticity per chain in loop structures larger than the smallest.

Perfect Network Formation

The classical Flory-Stockmayer (F-S) treatment of the gel point and the accompanying changes in distributions of molecular species give a basic explanation of the phenomena to which the behaviour and changes in actual polymerisations may be related. However, as discussed in detail by Flory[1], the infinite species which occur from the gel point to complete reaction cannot be enumerated as individual molecules. In addition, F-S theory says nothing concerning the detailed topology of the network, which grows and defines its structure through the random reaction of its reactive groups with other groups on the gel and with groups on sol species. To obtain a perfect network, all reactions (sol-sol, sol-gel and gel-gel) are assumed to yield elastically active chains between junction points in the final network. This assumption is rarely true and will be examined in detail in this paper.

The elastomeric properties of polymer networks depend to a large extent on the value of the molar mass of the elastically active chain connecting a pair of junction points, M_c. If a perfect network structure is assumed then M_c can be calculated directly from the reactant structures, taking account of unreacted groups for non-stoichiometric reaction mixtures. In general, the

© WILEY-VCH Verlag GmbH, D-69469 Weinheim, 2000

detailed relationship between M_c and reactant structures depends on reactant architectures as well as extents of reaction and it is not possible to give a completely general formula. For example, comb-like reactants with pendant reactive groups give loose ends, which are connected to only single branch points, and two types of chains between junction points, those involving sections of the backbone chains and those involving side chains. However, formulae for M_c from most stoichiometric endlinking polymerisations (using star reactants) at complete reaction can be derived relatively simply by assuming the perfect network structure and relating it to the structures of the reactants from which it is formed[2,3]. In such networks there are no loose ends and M_c is related to the molar concentration and number of moles of chains connecting pairs of junction points, n_c and N_c, respectively, by the relationships

$$n_c = \frac{N_c}{V_{net}} = \frac{N_c}{W_{net}} \cdot \frac{W_{net}}{V_{net}} = \frac{1}{M_c} \cdot \rho ,$$ (1)

where W_{net} and V_{net} are the mass and the volume of network, respectively, and ρ is the density of the network. It is assumed that any sol fraction has been removed. $1/M_c$ for networks formed from several reactants is an average value, denoted $<1/M_c>$, and may be related to reactant structures after introduction of N_J, the number of junction points in the network, to give

$$< \frac{1}{M_c} > = \frac{N_c}{N_J} \cdot \frac{N_J}{W_{net}} .$$ (2)

General expressions for $<1/M_c>$ may be derived[2,3] for quite complex polymerisations, e.g. $\Sigma_i RA_{fai} + \Sigma_j R'B_{fbj}$ polymerisations. However, for present purposes, considerations are limited to $RA_2 + R'B_f$ polymerisations. In that case,

$$\frac{N_c}{N_J} = \frac{f}{2} ,$$ (3)

as $f/2$ chains emanate from each junction and M_c ($= W_{net}/N_c$) is simply the molar mass of two arms of the $R'B_f$ unit plus the molar mass of the RA_2 unit.

Relationships between concentrations of chains and junction points that assume perfect network structures, such as equations (1) to (3), are often used when interpreting elastic properties of endlinked networks[4]. All chains and junction points are assumed to be elastically active. In this case, one may replace the symbols N_c and N_J by N_{ec} and N_{eJ}, denoting the numbers of elastically active chains and junction points. In practice, deviations from these assumptions occur and the values of N_{ec} or N_{eJ} deduced from elastomeric properties are rarely those expected from the amount of chemical reaction that has occurred. Such deviations may be due to topological entanglements and chain interactions[5-9], to side reactions, incomplete reaction in endlinking polymerisations (giving loose ends)[10,11] and, more fundamentally and generally, inelastic chain or loop formation due to the intramolecular reaction of pairs of groups[2,3,9].

It is obviously important for predicting and interpreting the elastomeric properties of networks to be able to calculate the value of N_{ec}, N_{eJ} or M_c from the reactants and the reaction conditions used. For stoichiometric $RA_2 + R'B_f$ polymerisations at complete reaction, one may write M_c as

$$M_c^o = \frac{W_{net}}{N_{ec}^o} \quad , \tag{4}$$

where M_c^o and N_{ec}^o refer to the perfect network structure. In reality, due to intramolecular reaction the actual number of elastic chains, N_{ec}, is less than N_{ec}^o, giving

$$M_c = \frac{W_{net}}{N_{ec}} \quad , \tag{5}$$

with $M_c > M_c^o$. Thus, M_c may be considered as the key quantity to consider, both experimentally and theoretically.

Elastic Modulus

The close connection between network structures and modulus (G) is summarised by the equations, based on the application of Gaussian elasticity theory to uniaxial deformation[2-4],

$$\sigma = G(\Lambda - \Lambda^{-2}) \tag{6}$$

and

$$G = ART\rho\varphi_2^{1/3}(V_u / V_F)^{2/3} / M_c \quad . \tag{7}$$

Equations (6) and (7) are valid at relatively low deformations. G is the shear modulus measured at zero frequency. σ is the applied stress per undeformed area and Λ the experimental deformation ratio. In the absence of free chain-ends, A has the values of $(1 - 2/f)$ and 1 for phantom and affine chain behaviours, respectively. Also, A may be put equal to 1 for values of Λ near 1. ρ is the density of the dry network, ϕ_2 the volume fraction of network during measurement ($\phi_2=1$ for unswollen networks having no sol fraction), V_u is the volume of the dry, unstrained network and V_F the volume at formation (assumed to be equivalent to V_o, the volume in the strain-free reference state.) Hence, values of M_c may be deduced from measurements of modulus using equation (7). For dry networks, prepared in bulk $V_u = V_F$, ϕ_2 = 1 and, for measurements at small deformations, $A = 1$, thus

$$G = \frac{\rho}{M_c} \cdot RT = \frac{N_{ec}}{V_{ec}} \cdot RT = n_{ec} RT . \tag{8}$$

Intramolecular Reaction

In network-forming polymerisations, the increasing numbers of reactive groups per molecule, together with the spatial correlations between groups on the same molecule, mean that intramolecular reaction *cannot* generally be neglected. For -A + B- polymerisations, intramolecular reaction can be characterised in terms of the parameter P_{ab} for the smallest loop[2,3,12,13)]

$$P_{ab} = \frac{P(\underline{r} = \underline{0})}{N_{Av}} , \tag{9}$$

where $P(\underline{r} = \underline{0})$ is the probability-density of a zero end-to-end vector between reactive groups. P_{ab} thus represents the mutual concentration of A- and B-groups at the ends of the shortest sub-chain that can react intramolecularly. The structure of this sub-chain, consisting of ν skeletal bonds, and of root-mean-square end-to-end distance $\langle r^2 \rangle^{1/2}$, is shown in figure

1. If it assumed that the end-to-end distance distribution can be represented by a Gaussian function, P_{ab} is given by

$$P_{ab} = \frac{1}{N_{Av}} \left\{ \frac{3}{2\pi < r^2 >} \right\}^{3/2} . \tag{10}$$

Since the units of P_{ab} are moles per unit volume, it can be described as the mutual *concentration* of a pair of reactive groups on the same molecule that can react intramolecularly, and is said [2,3] to define

an *internal* concentration, c_{int}. The concept is shown schematically in figure 2 for an A-group, where the competition between intramolecular and intermolecular reaction

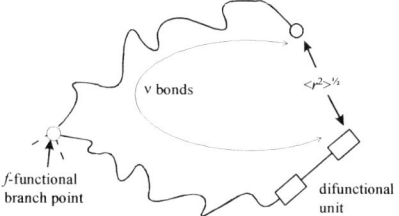

Fig. 1: sub-chain forming a smallest loop structure illustrated with respect to an RA$_2$+RB'$_f$ polymerisation. The diagram shows the two arms of a star reactant, one arm having reacted with a difunctional monomer; the root-mean-square distance of the chain of ν bonds between terminal groups is $\langle r^2 \rangle^{1/2}$.

with B-groups is due to the (internal) concentration of B-groups on the same molecule (described by $c_{b,int}$) relative to that of B-groups on other molecules, the so-called external concentration, $c_{b,ext}$. The latter quantity may be approximated by the simple expression

$$c_{b,ext} = c_b = c_{b0}(1 - p_b) , \tag{11}$$

where c_b is the instantaneous concentration of reactive groups (at extent of reaction p_b). A corresponding expression exists for the intermolecular reaction of a B-group, namely

$$c_{a,ext} = c_a = c_{a0}(1 - p_a) .$$ (12)

Intramolecular reaction between the pendant A- and B-groups at the ends of the sub-chain shown in Fig. 1 results in the formation of the smallest loop. However, it is apparent that loop sizes, consisting of larger numbers of monomer (reactant) units are possible, resulting in ring structures with integer multiples of v bonds. If all such sub-chains in a polymerisation mixture are assumed to obey Gaussian statistics, the parameter characterising the formation of the ith size of loop can be evaluated using a more general form of equation (10),

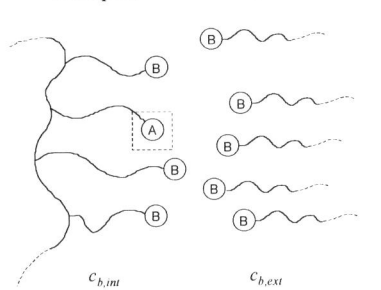

$$P_{ab,i} = \left(\frac{3}{2\pi i <r^2>}\right)^{\frac{3}{2}} \frac{1}{N_{Av}} = P_{ab}.i^{-\frac{3}{2}} .$$ (13)

It is assumed that the presence of i branch points in a sub-chain forming a loop of size i does not disproportionately affect the conformational statistics.

Fig. 2: Illustrating the concepts of internal and external concentrations of B-groups around a chosen A-group, leading, respectively, to intramolecular and intermolecular reaction

A useful measure of the propensity of a system at a given ratio of reactants for intramolecular reaction is λ_{a0}, where

$$\lambda_{a0} = \frac{P_{ab}}{c_{a0}} ,$$ (14)

with c_{a0} the initial concentration of A-groups. λ_{a0} captures the combined effects of reactant structure and reactive-group concentration. A decrease in chain length or chain stiffness (i.e., a decrease in $<r^2>$) results in an increase in P_{ab} and, hence, in the probability of intramolecular reaction and the formation of loop structures. Similarly, decreasing the concentration of reactive groups enhances the probability of intramolecular reaction.

Experimental Results for Polyurethane (PU) Networks

The results to be discussed come from six series of PU-network materials formed via stoichiometric $RA_2 + R'B_f$ polymerisations of hexamethylene diisocyanate (HDI) and star polyoxypropylene (POP) polyols, at different initial dilutions in nitrobenzene[14,15]. The structure of the polyol (R'B$_f$) was varied in order to examine the effects of branch-point functionality (f) and reactant sub-chain length (ν) on the moduli of the resulting networks. The six reaction systems are listed in Table 1. As described in equation 4, M_c^o is the network-chain molar mass in the perfect network and is defined by the reactant structures. By rearranging the sub-chain illustrated in Fig. 1, it is easy to show that M_c^o is also the molar mass of the sub-chain of ν skeletal bonds[15]. The values of $\langle r^2 \rangle$ listed enable values of P_{ab} to be calculated using equation (10).

Table 1: Functionalities, f, numbers of skeletal bonds, ν, in the sub-chains forming the smallest possible loops and elastic chain molar masses in the perfect networks, M_c^o, for six series of stoichiometric PU-forming, nonlinear polymerisations[15]. The calculated values, using detailed conformational analyses[16], of the mean-square end-to-end distances, $\langle r^2 \rangle$, of the sub-chains of ν bonds embedded in branched structures are also shown.

PU system	f	ν	M_c^o/g mol^{-1}	$\langle r^2 \rangle$ /nm^2
1. HDI + POP triol	3	35	635	3.718
2. HDI + POP triol	3	62	1168	6.877
3. HDI + POP tetrol	4	28	500	2.753
4. HDI + POP tetrol	4	32	586	3.628
5. HDI + POP tetrol	4	43	789	4.605
6. HDI + POP tetrol	4	65	1220	6.581

The plots in Fig. 3 give the values of M_c, relative to M_c^o for the perfect networks (M_c/M_c^o), as functions of the average initial dilution of reactive groups, $2/(ca_0+cb_0)$, for the six series of PU networks formed at complete reaction. The values of M_c were determined from uniaxial compression measurements on dry and swollen networks and analysis of the results using equations (6) and (7). For each of the six experimental systems, an increase in the initial dilution of reactive groups results in greater reductions in moduli, consistent with an increased incidence of intramolecular reaction, and the formation of inelastic loop structures. The positive slopes of the plots indicate that direct relationships exist between intramolecular reaction (which increases with reactant dilution) and the network defects at complete reaction. This in itself shows that the dominant network defects are inelastic loop structures, which can form both pre-gel and post-gel.

The plots in Fig. 3 clearly show that the magnitudes of the experimentally observed reductions in modulus are by no means insignificant. If perfect networks were formed, then

$M_c / M_c^o = 1$. Hence, there is a ten-fold decrease in modulus in the case of the dry (bulk) network formed from the short-chain triol (system 1; $M_c^o = 635$ g mol^{-1}), at the highest initial

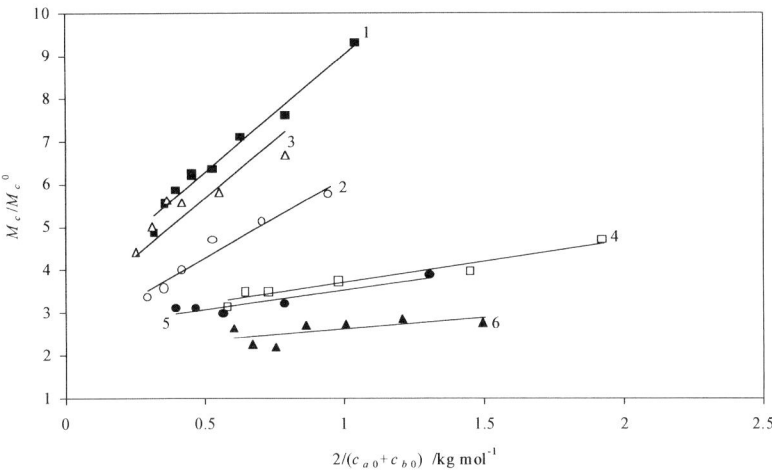

Fig. 3: Experimentally-determined values of M_c / M_c^o at complete reaction as functions of the average initial dilution of reactive groups, $2/(c_{a0}+c_{b0})$, for the six series of PU networks of Table 1.

dilution of reactive groups. For a given branch-point functionality, an increase in sub-chain length (v) results in a decrease in M_c / M_c^o, due to the decrease in the probability of loop formation (since $<r^2>$ in equation (13) increases). At approximately the same values of v, the reductions in moduli for tetrafunctional networks are considerably less than those of trifunctional ones. To a first approximation, this can be understood [9], on the basis of the different effects of the smallest loops that can form during $f = 3$ and $f = 4$ polymerisations.

Also, it should be noted that if the *perfect* networks were to exhibit phantom rather than affine behaviour, then $A = (1 - 2/f)$ (equation (7)) and for $f = 3$, $M_c / AM_c^o = 3$, and for $f = 4$, $M_c / AM_c^o = 2$. The observed values of M_c / M_c^o (or M_c / AM_c^o for any value of A) are greater, therefore, than those for perfect networks assuming either affine *or* phantom behaviour.

General Effects of Loops on Elasticity

The effect of the smallest and next smallest loop structures ($i = 1,2$ in equation (13)) on the loss of network elasticity is illustrated in figure 4 for trifunctional and tetrafunctional networks. In perfect f-functional networks, each junction provides $f/2$ elastic chains. In the case of $f = 3$, each smallest loop structure renders two branch points inelastic (shown as $\square$ in the diagram), which is equivalent to the loss of three elastic chains. For $f = 4$, the effect is reduced; each loop is associated with the loss of a single elastic branch point, or two elastic chains. As a result, trifunctional networks are expected to be the more sensitive to loop defects, and decreases in modulus are expected to be greater than those of tetrafunctional networks with similar values of M_c^o, consistent with the experimental data in Fig. 3.

Unlike the smallest loops, larger loop structures do not disrupt the continuity of the network structure, and therefore network chains in larger loops *are* capable of supporting a load. The examples of the two-membered loop structures for trifunctional and tetrafunctional networks are also shown in figure 3. The question as to *how* larger loop-structures contribute to losses in network elasticity remains unanswered. The conformational entropy of a large loop-structure will be reduced[3,17,18], relative to that of an unperturbed, free linear chain of the same number of skeletal bonds, due to the decrease in the total number of possible

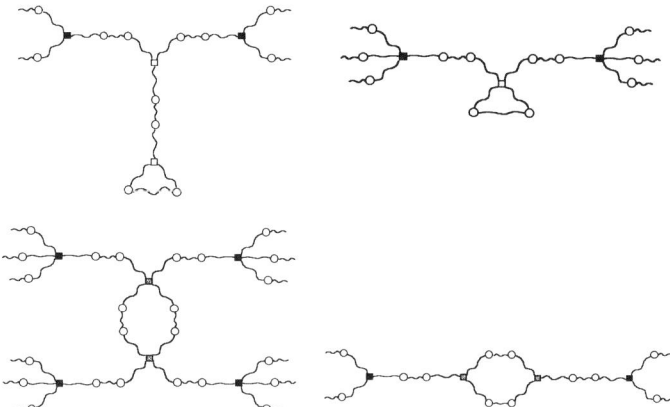

Fig. 4: Smallest (one-membered) and next smallest (two-membered) loop structures ($i = 1,2$ in equation (13)) in trifunctional and tetrafunctional networks. $\bigcirc$ represents a reacted pair of groups; $\blacksquare$ denotes a fully-elastic junction point; $\blacksquare$ denotes a junction point of reduced elasticity; $\square$ denotes an elastically-inactive junction point.

chain conformations resulting from the constraints imposed by the branch points along the chain and at the two chain ends. However, in a *completely reacted* network structure, *every* chain (except those in smallest loops) must form part of a topological circuit, and will therefore be subject to some degree of entropy loss. Since the origin of rubber-like elasticity lies in the conformational entropy of the network chains, any decrease in entropy should manifest itself as a decrease in the elasticity of the real network structure relative to that of the *hypothetical*, perfect one, whose network chains are assumed to be indistinguishable from the corresponding set of unperturbed (free) chains.

Monte-Carlo Polymerisation Algorithm

Theories to predict the modulus of a network material must begin by constructing a realistic model of the network structure, including defects[17,19-21]. Detailed characterisation of the connectivity, or topology, of a network structure by conventional, experimental means is impossible. In order to investigate the effects of network topology on elastomeric properties one must therefore use numerical simulations of the network-forming nonlinear polymerisations. These have the potential to provide this detailed structural information. In such simulations, it is important to account correctly for the formation of loop-structures of various sizes resulting from intramolecular reactions correctly weighted according to their probabilities of formation. To these ends, a Monte-Carlo (M-C) nonlinear polymerisation algorithm, originally devised by Dutton, Stepto and Taylor[17] has been further developed to simulate self-polymerisations (RA_f), and two-monomer polymerisations of the general type $RA_2+R'B_f$. During the course of a simulated polymerisation, populations of monomer units are connected together according to the relative probabilities for intramolecular and intermolecular reactions using λ_{a0} and taking account of possible loop sizes and the decreasing external concentration of reactive groups as a polymerisation proceeds. All the connections are recorded as a function of extent of reaction of A- or B-groups, along with the calculated sol and gel fractions, and average degrees of polymerisation.

The algorithm requires an initial number of A- and B-bearing reactants, $N_a + N_b$ (where $n_{a0} = f_a N_a = n_{b0} = f_b N_b$, for stoichiometric reactions), with functionalities of $f_a = 2$ and $f_b = 3,4$ for the systems to be discussed here. The initial concentrations of reactive groups, c_{a0} and c_{b0}, are also specified. These effectively define a reaction volume for given reactant molar masses. The final parameter characterising the reactants is P_{ab}, so that λ_{a0} (equation (14)) is known.

Loop-Size Distributions and Extents of Intramolecular Reaction

The general description and operation of the algorithm have been described elsewhere[21]. Here, we concentrate on findings relevant to the experimental results in Fig. 3. Accordingly, Fig. 5 shows distributions at complete reaction for an F-S polymerisation and polymerisations with $\lambda_{a0} = 0.01$ and 0.1. The number fraction of loop structures of i repeat units, $n_r(i)$, is plotted against i.

$$n_r(i) = \frac{N_r(i)}{\sum_i N_r(i)} \quad , \tag{15}$$

where $N_r(i)$ denotes the number of loop structures of size i. The F-S simulation was achieved by modifying the algorithm to exclude sol-sol and sol-gel intramolecular reaction and to allow only random intramolecular reaction on the largest species beyond the F-S gel point, i.e., random gel-gel reaction. Smallest loops are negligible under F-S conditions. However, it is clearly seen that smallest loops start to dominate as λ_{a0} increases. Maxima in the distributions reflect the different numbers of opportunities for forming loop structures of various sizes, integrated over all post-gel reaction. Also, the maximum is shifted to slightly smaller loop sizes in the simulations using λ_{a0} compared with those using F-S statistics. This is to be expected since intramolecular reactions are then biased towards the formation of smaller loops.

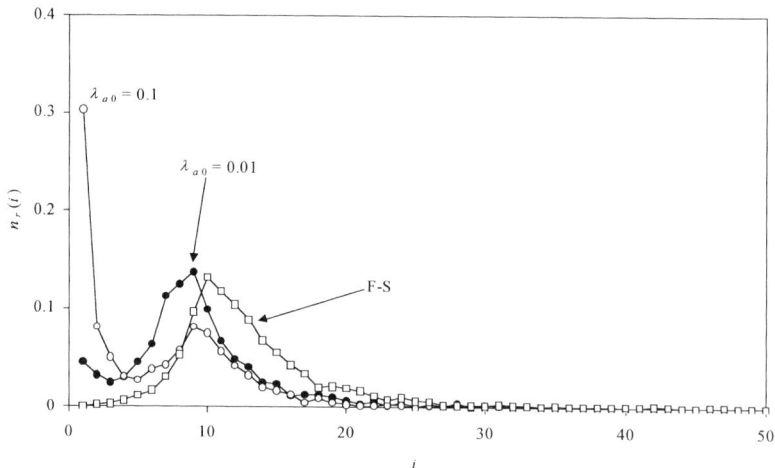

Fig. 5: $RA_2+R'B_4$ simulations at complete reaction. Overall number-fraction loop-size distributions for F-S, $\lambda_{a0} = 0.01$ and $\lambda_{a0} = 0.01$ simulations, each containing 1000 branch units.

The changes in the calculated loop-size distributions in the *gel molecule* at complete reaction with increasing λ_{a0} can be simply characterised by calculating extents of intramolecular reaction on that molecule resulting in smallest loops by the end (*e*) of a polymerisation, $p_{re,1}$, as a function of λ_{a0}. Such a characterisation is useful because the exact effects of smallest loops on network elasticity can be deduced (see Fig. 4). In addition, $p_{re,i>1}$, the extent of reaction resulting in the formation of *larger* loops may also be calculated, so that the total, final extent of intramolecular reaction may be written

$$p_{re,tatal} = p_{re,1} + p_{re,l>1} \; . \tag{16}$$

Accordingly, series of stoichiometric $RA_2+R'B_3$ and $RA_2+R'B_4$ simulations with 1000 branch units were performed with λ_{a0} in the range 0.01 to 0.40, covering the experimental polymerisations in Fig. 3. For illustration, the results for the series of $RA_2+R'B_3$ simulations are shown in Fig. 6. Notice that $p_{re,tatal} = 1/6$, always. This is related to the cycle rank of the completed network. For a stoichiometric $RA_2+R'B_f$ polymerisation at complete reaction[22] $p_{re,tatal} = (1/2)(1 - 2/f)$.

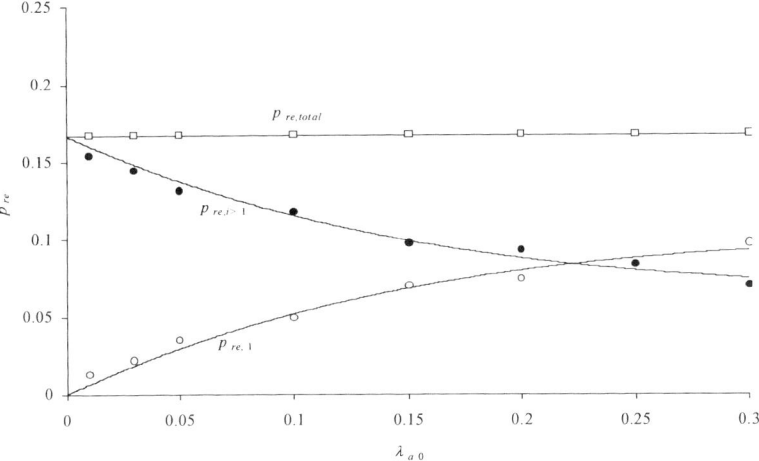

Fig. 6: $p_{re,1}$ and $p_{re,i>1}$ versus λ_{a0} for $RA_2+R'B_3$ simulations with 1000 branch units.

Fig. 6 shows that the incidence of smallest loops ($p_{re,1}$) is negligible when $\lambda_{a0} = 0$. Essentially, in agreement with F-S statistics and the results in Fig. 5, all intramolecular reaction then occurs *after* the gel point, resulting in larger loop structures. However, as λ_{a0} increases, $p_{re,1}$ increases, at the expense of the proportion of larger loop structures ($p_{re,i>1}$). For a given value of λ_{a0}, it is found that tetrafunctional systems give rise to more loop structures than trifunctional systems, simply due to the greater number of opportunities for loop-formation in the former case. However, relative to the *total* number of loop structures, the proportion of smallest loops appears to be fairly insensitive to the branch-point functionality[21].

Correlation of Model Network Topologies with Measured Network Moduli

The effects of loop structures on experimental reductions in moduli can be taken into account, approximately, by basing the numbers of elastically effective chains on values corresponding to $p_{re,1}$ and $p_{re,i>1}$ from the M-C simulations. If the number of chains rendered elastically ineffective, N_c^I, can be estimated, the reduction in modulus, or increase in M_c, can be calculated using

$$\frac{M_c}{M_c^o} = \frac{N_{ec}^o}{N_{ec}} = \frac{N_{ec}^o}{N_{ec}^o - N_c^I} \quad . \tag{17}$$

From equation 3, $N_{ec}^o = N_J \cdot f / 2$ and the quantity N_c^I may be estimated as follows. For the simulated trifunctional networks of N_J branch units, the numbers of smallest loops are given by $3N_J p_{re,1}$, which equates to a loss of elastic chains equal to $3 \times 3N_J p_{re,1}$. However, the $3 \times 3N_J p_{re,i>1}$ chains in *larger* loops are subject only to a partial loss in elasticity, and may be taken as equivalent to $x \times 3 \times 3N_J p_{re,i>1}$ chains, where x is the fractional loss in elasticity per chain in a larger loop structure. ($x = 1$ corresponds to a *total* loss in elasticity, seen in the case of smallest loops only.) Hence, for $f = 3$

$$N_c^I = \frac{3N_J}{2} \cdot \left(6p_{re,1} + x \cdot 6p_{re,i>1}\right) \tag{18}$$

and substituting for N_c^I and N_{ec}^o in equation (17) yields

$$\frac{M_c}{M_c^o} = \frac{1}{1 - 6p_{re,1} - x \cdot 6p_{re,i>1}} \quad . \tag{19}$$

A similar expression can be derived for tetrafunctional network structures. They lose only 2 elastic chains per smallest loop and

$$\frac{M_c}{M_c^o} = \frac{1}{1 - 4p_{re,1} - x \cdot 4p_{re,i>1}} \quad . \tag{20}$$

Estimates of x, the fractional loss of elasticity for chains in larger loop structures, can be made using the experimentally determined values of M_c / M_c^o for the series of PU networks in Fig. 3, in conjunction with simulated values of $p_{re,1}$ and $p_{re,i>1}$ and equations (19) and (20). However, since $p_{re,1}$ and $p_{re,i>1}$ calculated via the M-C simulations are dependent upon P_{ab}, values of M_c / M_c^o depend on both x and P_{ab}. Correlations between M-C calculations and experimental data may therefore be performed in two ways:

i) bivariate least-squares fitting, to evaluate both x and P_{ab}

ii) monovariate least-squares fitting, to evaluate x using P_{ab} values calculated *ab initio*, via chain-conformational analyses[16] and the values of $<r^2>$ in Table 1.

The calculated values of x and P_{ab} for PU systems 1 to 6 are listed in Table 2 and the fittings of the experimental M_c / M_c^o data using the bivariate and monovariate analyses are shown, respectively, in Figs. 7 and 8. The bivariate analysis results in very good fits to the

Table 2: Values of x and P_{ab} based on correlations of experimentally-measured reductions in moduli for PU networks, and extents of intramolecular reaction calculated from M-C simulations. [a]bivariate fitting; [b]monovariate fitting with P_{ab} values calculated *ab initio*, via chain-conformational analyses using $<r^2>$ values from Table 1.

PU system	F	x^a	$P_{ab}{}^a$ /mol l^{-1}	x^b	$P_{ab}{}^b$ /mol l^{-1}
1	3	0.667	0.538	0.816	0.076
2	3	0.599	0.103	0.690	0.030
3	4	0.684	0.547	0.775	0.120
4	4	0.638	0.457	0.730	0.079
5	4	0.582	0.129	0.637	0.055
6	4	0.481	0.088	0.559	0.032

experimental data and values of x in the range 0.67 to 0.60 are required to reproduce the experimental modulus reductions for the trifunctional PU networks (systems 1 and 2), and 0.68 to 0.48 for the tetrafunctional PU networks (systems 3 to 6). In both cases, an increase in the size of the smallest loop structure results in a decrease in x, indicating less elasticity lost. However, the values of P_{ab} estimated from the bivariate fitting are *much higher* than those calculated via $<r^2>$ for the sub-chain structures. Use of the *ab initio* calculated values of P_{ab} in the monovariate analysis results in a worse fitting of the experimental data at the higher values of M_c / M_c^o, but acceptable values of x are still required (x=0.82 - 0.69, for $f = 3$, and 0.78 - 0.56 for $f = 4$).

Finally, Fig. 9 illustrates the effects of assuming that $x = 0$ (*i.e.* no additional loss in elasticity from larger loop structures). The experimental data from system 3 are shown in comparison with the results from the bivariate and monovariate analyses and the monovariate analysis with $x = 0$. It is obvious that the neglect of the loss of elasticity in larger loop structures results in a *gross underestimation* of the experimentally observed reductions in moduli. The effects of the values used for P_{ab} are secondary compared with the effects of assuming $x = 0$.

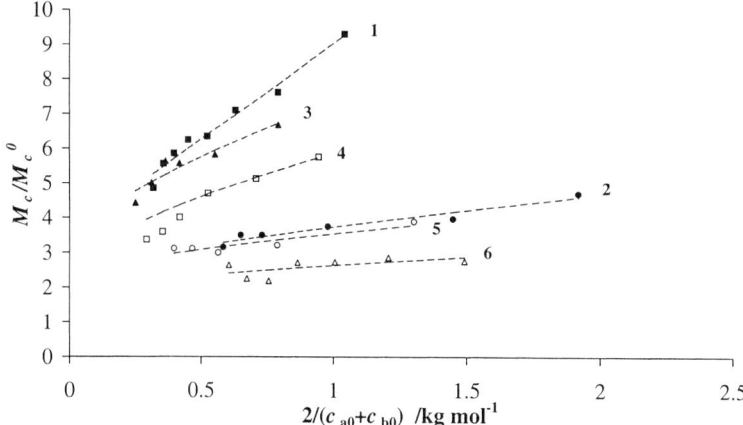

Fig. 7: M_c / M_c^o versus average initial dilution of reactive groups, $2/(c_{a0}+c_{b0})$, for the PU systems of Table 1 and Fig. 3. Experimental values of M_c / M_c^o and calculated values using the bivariate analysis.

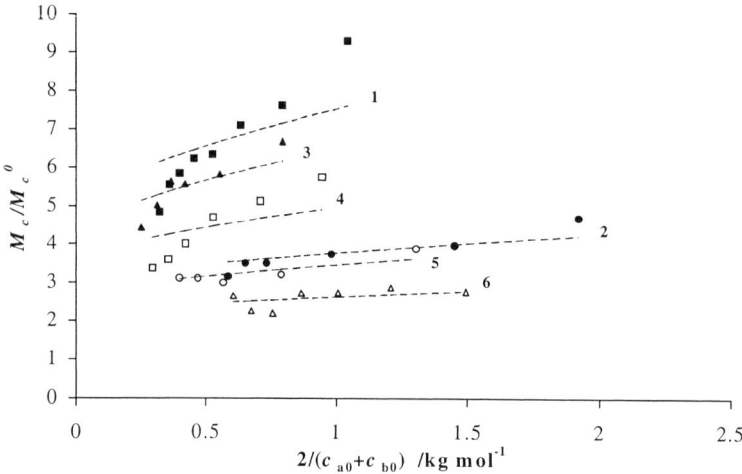

Fig. 8: M_c / M_c^o versus average initial dilution of reactive groups, $2/(c_{a0}+c_{b0})$, for the PU systems of Table 1 and Fig. 3. Experimental values of M_c / M_c^o and calculated values using the monovariate analysis.

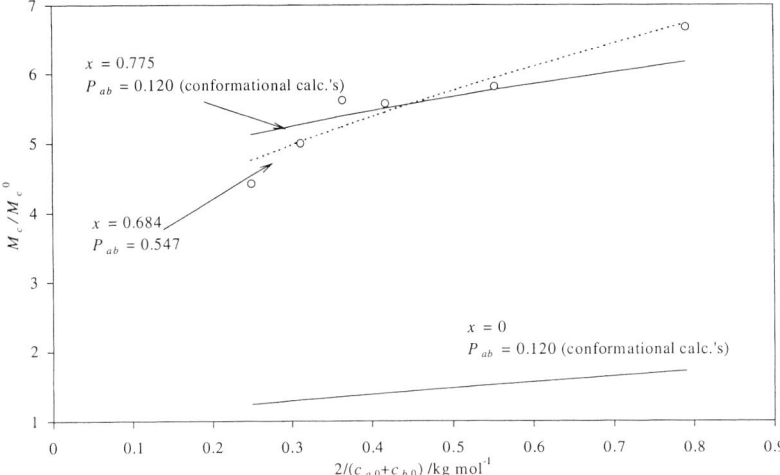

Fig. 9: M_c / M_c^o versus average initial dilution of reactive groups, $2/(c_{a0}+c_{b0})$, for PU system 3 of Table 1 and Fig. 3. Experimental values of M_c / M_c^o and calculated values using the bivariate and monovariate analyses and the monovariate analysis with $x = 0$.

Discussion and Conclusions

Although the incidence of unreacted chain ends in a network material can, in principle, be quantified experimentally, enumerating network defects due to loop structures remains experimentally intractable. An M-C simulation approach, in which the numbers of loop structures formed are weighted according to size and the molecular structures of the reactants, can provide this detailed information, providing a means of generating realistic model network structures, whose topologies are known in full detail.

The correlations described, between experimentally measured network moduli and extents of loop-forming reaction calculated via M-C simulation, show clearly that larger loop structures contribute significantly to the observed loss of elasticity relative to that of a perfect network.Current work is focusing on the direct calculation of the entropies of larger loop structures and more exact calculations of P_{ab} for sub-chains in branched structures. From such calculations, moduli can be predicted more accurately, *directly* from reactant structures and reaction conditions, without the use of semi-empirical value of x and P_{ab}.

Acknowledgement

Support of the EPSRC for grant GR/L/66649 is gratefully acknowledged.

References

1. P.J. Flory, *"Principles of Polymer Chemistry"*, Cornell University Press, Ithaca 1953
2. R.F.T. Stepto, in *"Comprehensive Polymer Science"*, 1st Suppl., S.L. Aggarwal and S. Russo, eds., Pergamon Press, Oxford, 1992, chap. 10
3. R.F.T. Stepto, in *"Polymer Networks – Principles of Their Formation, Structure and Properties"*, R.F.T. Stepto, ed., Blackie Academic & Professional, London, 1998, chap. 2
4. J.E. Mark and B. Erman, in *"Polymer Networks – Principles of Their Formation, Structure and Properties"*, R.F.T. Stepto, ed., Blackie Academic & Professional, London, 1998, chap. 7
5. W.W. Graessley, *Adv. Polymer Sci.*, **16** (1974)
6. M. Doi and S.F. Edwards, *J. Chem. Soc., Faraday Trans. 2*, **74**, 1789, 1802, 1818 (1978)
7. G. Heinrich, E. Straube and G. Helmis, *Adv. Polymer Sci.*, **85**, 34 (1988)
8. O. Kramer, in *"Elastomeric Polymer Networks"*, J.E. Mark and B. Erman, eds., Prentice Hall, Englewood Cliffs, NJ, 1992, chap. 17
9. R.F.T. Stepto and B.E. Eichinger, in *"Elastomeric Polymer Networks"*, J.E. Mark and B. Erman, eds., Prentice Hall, Englewood Cliffs, NJ, 1992, chap. 18
10. M. Gottlieb, C.W. Macosko, G.S. Benjamin, K.O. Meyers and E.W. Merrill, *Macromolecules*, **14**, 1039 (1981)
11. C.W. Macosko and J.C. Saam, *Polymer. Prepr., Div. Polymer Chem., Amer. Chem. Soc.*, **26**, 48 (1985)
12. H. Jacobson and W.H. Stockmayer, *J. Chem. Phys.*, **18**, 1600 (1950)
13. S.B. Ross-Murphy and R.F.T. Stepto, in *"Large Ring Molecules"*, J.A. Semlyen, ed., Wiley, Chichester, 1996, chap.16
14. J.L. Stanford and R.F.T. Stepto, in *"Elastomers and Rubber Elasticity, Amer. Chem. Soc. Symp. Series 193"*, J.E. Mark and J. Lal, eds., American Chemical Society, Washington D.C., 1982, chap. 20
15. R.F.T. Stepto, in *"Biological and Synthetic Polymer Networks"*, O. Kramer, ed., Elsevier Applied Science, Barking, 1988, chap. 10
16. D.J.R. Taylor and R.F.T. Stepto, *to be published*
17. S. Dutton, R.F.T. Stepto and D.J.R. Taylor, *Angew. Makromol. Chem.*, **240**, 39 (1996)
18. W.W. Graessley, *Macromolecules*, **8**(2), 186, 865 (1975)
19. K.-J. Lee and B.E.Eichinger, *Polymer*, **31**, 406, 414 (1990)
20. E.S. Page and L.B. Wilson, *"An Introduction to Computational Combinatorics"*, Cambridge University Press,1979, p.74 *et seq*
21. R.F.T. Stepto and D.J.R. Taylor, in *"Cyclic Polymers"*, 2nd ed., J.A. Semlyen, ed., Kluwer, 1999, *in press*
22. R.F.T. Stepto, *Polym. Bull. (Berlin)*, **24**, 53 (1990)

Macromol. Symp. 159, 179–186 (2000)

Poly(N-vinylcaprolactam) Microgels and its Related Composites

Shufu Peng[1] and Chi Wu,[1,2,*]

1. Department of Chemistry, The Chinese University of Hong Kong, Shatin, Hong Kong

2. The Open Laboratory of Bond-selective Chemistry, Department of Chemical Physics, University of Science and Technology of China, Hefei, Anhui, China

SUMMARY: Spherical hydrogels were prepared by precipitation copolymerization of N-vinylcaprolactam and sodium acrylate [P(VCL-co-NaA)] at 60 °C. As a thermally sensitive polymer, the increase of temperature in the range 25–40 °C can lead to a continuous shrinking of PVCL homopolymer chains. The copolymerization of a few molar percent of NaA into a PVCL chain increases its swelling extent and shifts its shrinking temperature slightly higher. Our results revealed that in the shrinking process, calcium ions (Ca^{2+}) induced a profound complexation of the P(VCL-co-NaA) microgels at a critical temperature (T_c) which was nearly independent of the NaA content. However, both the rate and degree of the complexation increase as the NaA content increases at T_c. One the basis of this study, we further studied the complexation of P(VCL-co-NaA) microgels with gelatin in the presence of calcium ions and prepared chemically crosslinked of the gelatin gels with embedded P(VCL-co-NaA) microgels. Our results showed that the shrinking of the embedded P(VCL-co-NaA) microgels could lead to the shrinking of gelatin gels. The biomedical application of this gelatin/microgel composite have been exploited.

Introduction

Recently, swelling and shrinking of hydrogels under the different conditions, such as temperature, pH, composition, ionic strength and solvent, have attracted much attention. For example, hundreds of experimental and theoretical studies on thermally sensitive poly(N-isopropylacrylamide) (PNIPAM) gels have been reported[1]. In general, a thermally sensitive hydrogel has a lower critical solution temperature (LCST); namely, it swells at lower temperatures but shrinks as the temperature increases. It is known that introducing a few molar percent of hydrophobic or hydrophilic groups into a hydrogel can alter its shrinking temperature and swelling extent[2]. In contrast, reports on thermally sensitive microgels are much limited, partially because their preparation and observation are relatively more difficult. Experimentally, it is important to prepare microgels with a sufficient amount of crosslinking points so that each microgel is still a swollen three-dimensional polymer network. Using microgels offers several advantages. For example, microgels can nearly instantly reach their

CCC 1022-1360/00/$ 17.50+.50/0

shrinking and swelling limits in comparison with days required by bulk gels. Microgels can also be injected in some special applications.

Poly(N-vinylcaprolactam) (PVCL) is a relatively new type of nonionic water-soluble polymer. It was developed for hair-care and cosmetic applications. In principle, it should be more biocompatible than PNIPAM. It has been known that PVCL can complex with organic compounds[3], resist hydrolysis, and its gel can undergo a continuous volume transition in the temperature range 25–36 °C[4]. Up to now, only a few studies on PVCL and its gels have been reported partially because its polymerization is more difficult and partially because its volume transition is not as sharp as that of the PNIPAM gels. However, its biocompability attracts us to initiate this study to see whether PVCL microgels can be used as an injecting composition for certain biomedical applications. It is also known that a certain kind of metal ions, such as Ca^{2+}, can interact with carboxylic groups on polymer chains via a polyion/metal complexation to form interchain aggregation[5]. This complexation led us to think whether it can be used to immobilize microgels incorporated with a proper amount of carboxylic groups inside body after injection. This study is a fundamental research with some envisioned biomedical applications.

Experimental

Materials. N-vinylcaprolactam monomer (VCL, courtesy of BASF) was further purified by a reduced pressure distillation. Sodium acrylate monomer (NaA, from Lancaster) was used without further purification. Potassium persulfate as an initiator (KPS, from Aldrich) and N',N'-methylenebisacrylamide as a cross-linking agent (MBAA, from Aldrich) were recrystallized three times in methanol. The gelatin sample (courtesy of BASF) had a weight average molar mass of 1.5×10^5. To have a complete dissolution of gelatin, 2% of formamide was added into water to break interchain hydrogen bonding. Calcium chloride (anhydrous $CaCl_2$, from ACROS) was used without further purification.

Microgel preparation. Spherical poly(N-vinylcaprolactam-co-sodium acrylate) [P(VCL-co-NaA)] microgels were prepared by precipitation polymerization in water. Into a 150-mL three-neck flask equipped with a reflux condenser, a thermometer and a nitrogen-bubbling tube, were added 7.3 mmol VCL monomer, a proper amount of NaA comonomer, 0.19 mmol MBAA, and 40 mL deionized water. The solution was stirred and bubbled by nitrogen for 1 hr to remove oxygen before 0.045 mmol KPS aqueous solution was added to start the polymerization at 60 °C for 24 hrs. The resultant P(VCL-*co*-NaA) microgels were purified by a successive four times centrifugation (Sigma 2K15 ultracentrifuge, at 15,300 rpm and 40 °C), decantation and redispersion in deionized water to remove unreacted low molar mass molecules. Such obtained P(VCL-*co*-NaA) microgels were diluted with deionized water to concentrations lower than $\sim 1 \times 10^{-5}$ g/mL in laser light scattering measurements. The microgels with different contents of NaA were labeled as P(VCL–*m*A), where "*m*" represents the average molar content of acrylic groups. Linear P(VCL-*co*-NaA) chains was prepared in a similar way without adding the crosslinking agent, MBAA.

Laser light scattering (LLS). The detail of our laser light scattering spectrometer can be found elsewhere[6]. In static LLS, the angular dependence of the absolute excess time-averaged scattered intensity, known as the Rayleigh ratio $R_{vv}(q)$, can lead to the weight-average molar mass (M_w), the z-average root-mean-square radius of gyration ($<R_g^2>_z^{1/2}$, or written as $<R_g>$) and the second virial coefficient (A_2), where q is the scattering vector. In dynamic LLS, the cumulant or Laplace inversion analysis of the measured intensity-intensity time correlation function $G^{(2)}(q,t)$ in the self-beating mode can result in an average line width ($<\Gamma>$) or a line width distribution ($G(\Gamma)$)[7]. For a pure diffusive relaxation, Γ is related to the translational diffusion coefficient D by $(\Gamma/q^2)_{C\to0,q\to0} = D$ and to the hydrodynamic radius (R_h) by the Stokes-Einstein equation, $D = k_B T/(6\pi\eta R_h)$, where k_B, T and η are the Boltzmann constant, the absolute temperature and the solvent viscosity, respectively[8].

Results and discussion

Table 1 summarizes static and dynamic LLS results of spherical microgels with different contents of NaA in the absence of Ca^{2+} in the swollen (27 °C) and collapsed (50 °C) states. In each case, M_w is independent of the temperature, indicating no interchain or no interparticle aggregation. On the other hand, the shrinking of the microgels as the temperature increases was relatively smooth, similar to the temperature-induced volume change of neutral PVCL microgels[4]. However, the ionic groups make the PVCL chain more hydrophilic, which leads to a higher swelling extent at lower temperatures, shifts the shrinking temperature higher, and results in a less compact globule[9]. The increase of the ratio of average gyration radius to average hydrodynamic radius ($<R_g>/<R_h>$) as the ionic content increases indicates further swelling of the microgels.

Figure 1 clearly shows the temperature dependence of the relative average hydrodynamic radius $\langle R_h \rangle/\langle R_h \rangle_{T=25°C}$ for P(VCL-*co*-NaA) microgels copolymerized with different amounts of NaA, where $\langle R_h \rangle_{T=25°C}$ is the average hydrodynamic radius $\langle R_h \rangle$ at 25° C. The relative shrinking of the microgels decreases as the ionic content increases. Note that in the temperature range studied, there was no change in M_w in each case, i.e., no inter-microgel aggregation in the shrinking process.

Table 1. Light scattering characterization of spherical P(VCL-*co*-NaA) microgels in water.

$\dfrac{T}{°C}$	P(VCL-*1.0*NaA)			P(VCL-*9.1*NaA)		
	$\dfrac{M_w \times 10^9}{g/mol}$	$\dfrac{\langle R_h \rangle}{nm}$	$\dfrac{\langle R_g \rangle}{\langle R_h \rangle}$	$\dfrac{M_w \times 10^9}{g/mol}$	$\dfrac{\langle R_h \rangle}{nm}$	$\dfrac{\langle R_g \rangle}{\langle R_h \rangle}$
27	1.34	330	0.70	5.54	515	1.29
37	1.40	191	0.73	5.41	371	0.91
50	1.42	134	0.85	5.21	272	1.11

Figure 2 shows a better view of the shrinkings of microgels in terms of the average chain density $(\langle \rho \rangle)$ defined as $M_w/((4/3)\pi \langle R_h \rangle^3)$. P(VCL-*1.0*NaA) microgels reached their corresponding collapsed states at 42 °C. The slow increase of $\langle \rho \rangle$ for the P(VCL-*9.1*NaA) microgels can be attributed to a balance between strong electrostatic repulsion and hydrophobic attraction. If using $\langle \rho \rangle$ at 25 °C as a reference, we found that $\langle \rho \rangle$ increases 8–16 times for the microgels, depending on the ionic content for the P(VCL-*1.0*NaA) microgels.

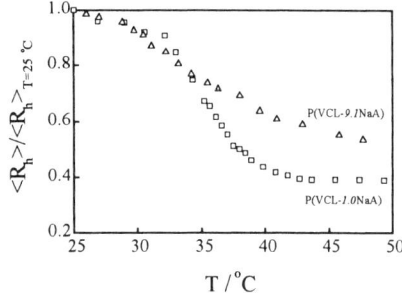
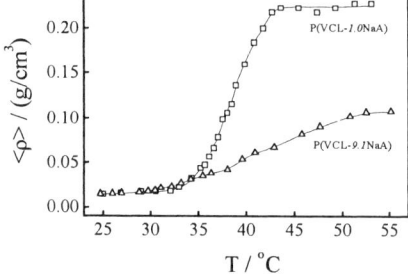

Fig. 1: Temperature dependence of the relativ average hydrodynamic radius $\langle R_h \rangle/\langle R_h \rangle_{T=25°}$ for P(VCL-*co*-NaA) microgels with differen amounts of NaA, where $\langle R_h \rangle_{T=25°C}$ is th average hydrodynamic radius $\langle R_h \rangle$ at 25° C.

Fig. 2: Temperature dependence of the average chain density $\langle \rho \rangle$ of P(VCL-*co*-NaA) microgels with different amounts of NaA.

Figures 3 and 4 show a complete different picture of the temperature dependence of the average hydrodynamic radius ($<R_h>$) and the apparent weight average molar mass ($M_{w,app}$) of microgels in a 0.03M $CaCl_2$ aqueous solution. In the range 25–31.8 °C, the microgel shrinks as the temperature increases, but $M_{w,app}$ is independent of the temperature, indicating no inter-microgel aggregation. At ~32 °C, the average hydrodynamic size and the apparent weight average molar mass sharply increases, revealing a clear inter-microgel aggregation. The aggregation number of the microgels (N_{agg}) varies from 40 to 200 as the ionic content increase. This is understandable because for the microgels, when the temperature reaches the transition temperature, PVCL becomes hydrophobic and starts to collapse, but the hydrophilic $-COO^-$ groups tend to stay on the periphery of the microgel. The complexation between Ca^{2+} ions and $-COO^-$ sticks the microgels together and the complexation occurred in a narrow temperature range. Our results showed that the resultant complexes have a similar average chain density $<\rho>$ of ~0.2 g/cm^3, in spite of a big different in N_{agg}. A slight decrease of $<\rho>$ as the ionic content increases can be attributed to the fact that the chains with more ionic groups are more hydrophilic and collapse less at high temperatures. Note that microgels have a amphiphilic character at higher temperatures. The shrinking of each linear chain forces most of the ionic groups to form a relatively more hydrophilic periphery, similar to a micelle[10], while in the case of the microgels, most of the ionic groups are trapped inside the microgel due to the crosslinking, so that there are less chances to form the inter-microgel complexes, which explains why N_{agg} (microgel) decreases as the ionic content increases.

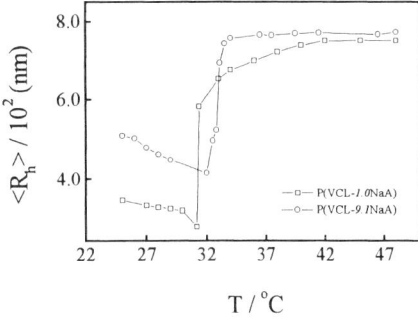

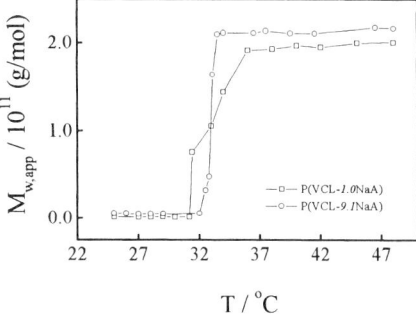

Fig. 3: Temperature dependence of the average hydrodynamic radius $<R_h>$ of spherical P(VCL-co-NaA) microgels in the presence of Ca^{2+}.

Fig. 4: Temperature dependence of the and the apparent weight average molar mass ($M_{w,app}$) of spherical P(VCL-co-NaA) microgels in the presence of Ca^{2+}.

184

Figures 5 and 6 show the temperature dependence of $<R_h>$ and $M_{w,app}$ for P(VCL-4.3NaA) microgels in the presence of different amounts of Ca^{2+}. Note that the complexation occurs at a similar temperature in spite of different Ca^{2+} concentrations, but both $<R_h>$ and $M_{w,app}$ increase as $[Ca^{2+}]$ increases. The increase of the ionic content has little effect on $<R_h>$. Our results reveal that the average number of the microgels inside each inter-microgel complex (N_{agg}) are ~60, ~20 and ~8 for $[Ca^{2+}] = 3 \times 10^{-2}$ M, 2×10^{-3} M and 2×10^{-4} M, respectively. The average chain density $<\rho>$ of the complexes decreases as N_{agg} decreases, which could be due to an imperfect packing of larger microgels inside each complexes, especially when each complex is only made of ~8 microgels.

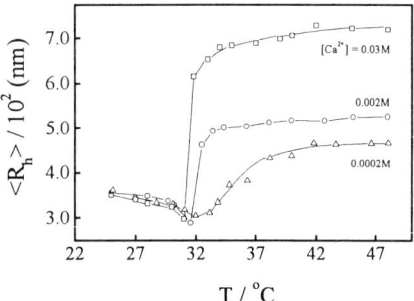

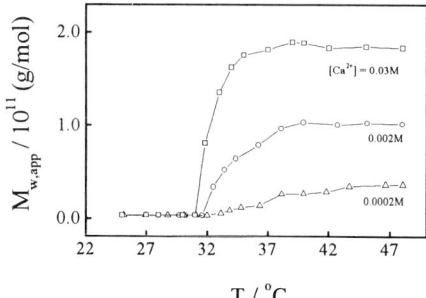

Fig. 5: Temperature dependence of the average hydrodynamic radius $<R_h>$ of P(VCL-4.3NaA) microgels with different Ca^{2+} concentrations.

Fig. 6: Temperature dependence of apparent weight average molar mass ($M_{w,app}$) of P(VCL-4.3NaA) microgels with different Ca^{2+} concentrations.

Figure 7 shows that temperature dependence of $<R_h>$ and $M_{w,app}$ of the P(VCL-4.3NaA)/gelatin mixtures with different gelatin/P(VCL-4.3NaA) ratios (C_g/C_p), where C_g and C_p are the average weight concentrations of gelatin and microgels, respectively. At T ~32 °C, both $<R_h>$ and $M_{w,app}$ of the mixtures slightly increase, revealing the mixtures undergo an association between the microgels and gelatin. A further increasing of the temperature led to a gradual decrease of $<R_h>$, but $M_{w,app}$ is independent of the temperature in the same range, indicating no further association. For the mixture with a lower ratio C_g/C_p, $<R_h>$ decreases smoothly as the temperature increases in the whole range studied. The results in Figure 7 indicates that the shrinking of the P(VCL-co-NaA) microgels could lead to the shrinking of the gelatin/microgel complexes. More P(VCL-co-NaA) microgels inside the complexes provide a larger shrinking force and a more compact final structure.

Figure 8 shows that the temperature dependence of $<R_h>$ and $M_{w,app}$ of P(VCL-4.3NaA)/gelatin mixtures in the presence of Ca^{2+}. Note that the complexation occurs at a similar temperature in spite of a difference in the gelatin concentration. The complexation in the presence of Ca^{2+} is much profound than in Figure 7. Both $<R_h>$ and $M_{w,app}$ increase as the C_g/C_p ratio decreases, indicating that gelatin acts as a stabilizer, presumably via an adsorption of gelatin chains on the microgels surface.

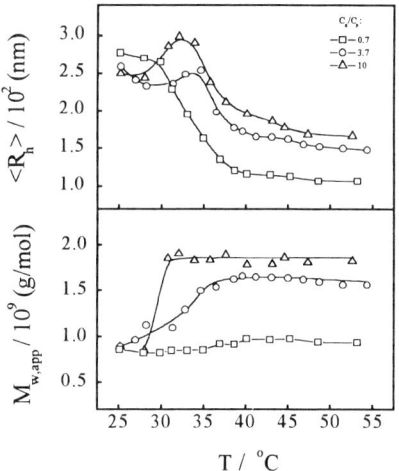 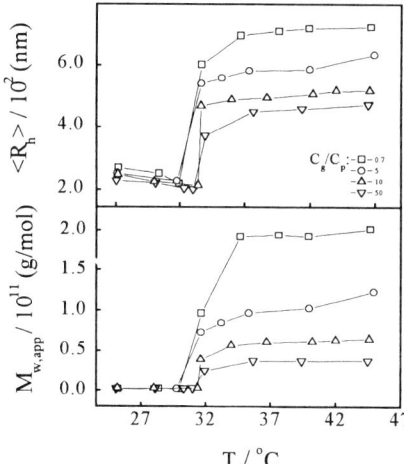

Fig. 7: Temperature dependence of the average hydrodynamic radius $<R_h>$ and apparent average molar mass weight average molar mass $(M_{w,app})$ of P(VCL-4.3NaA)/gelatin mixtures.

Fig. 8: Temperature dependence of the average hydrodynamic radius $<R_h>$ and the apparent weight average molar mass $(M_{w,app})$ of P(VCL-4.3NaA)/gelatin mixtures in the presence of Ca^{2+}.

Figure 9 shows that a bulk gel composite in which the microgels are embedded inside a chemically crosslinked gelatin network. It clearly shows that the shrinking of the microgels leads to a shrinking of the bulk gel composite. The degree of the shrinking can be controlled by the amount of the microgels embedded inside the gelatin network and by the crosslinking density. Such a gelatin/microgel composite is a potential biomedical material.

Figure 10 shows an *in vivo* biocompatibity test. The gel composite was inserted into a rat body at a sciatic never. The shrinking of a gel tube can connect a broken sciatic never at 37 °C. Note that the sciatic never has recovered after 20 weeks and the gel was completely absorbed by body.

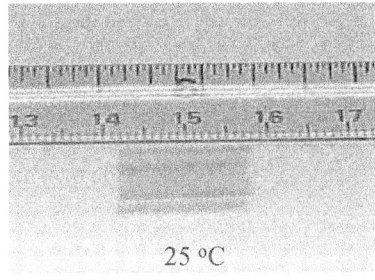

25 °C

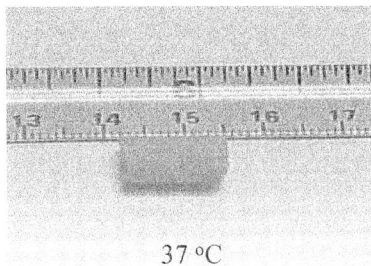

37 °C

Fig. 9: Swelling and shrinking of the PVCL/gelatin composite at the different temperatures.

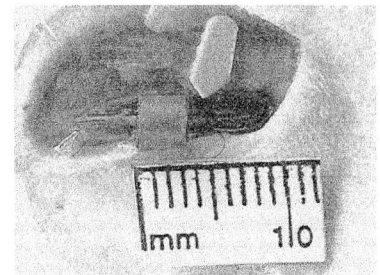

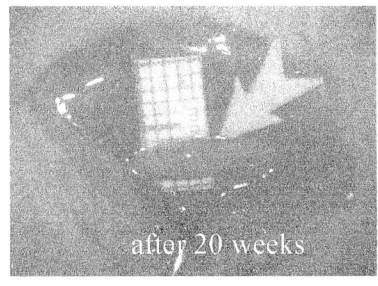

Fig. 10: Sciatic never before and after embedded the PVCL/gelatin gel tube into rat body.

References

1. S. Sat, M. Kong, H. Inmate, *Adv. Polym. Sci.* **109,** 207 (1993).
2. S. Zhou, B. Chu, *J. Phys. Chem. B* **102,** 1364 (1998).
3. M. Eisele, W. Burchard, *Makromol. Chem.* **191**, 169 (1991).
4. L. M. Mikheeva, N. V. Grinberg, A. Ya. Mashkevich, V. Ya. Grinberg, *Macromolecules* **30**, 2693 (1997).
5. S. Peng, C. Wu, *Macromolecules* **32**, 585 (1999).
6. X. Wang, Z. Xu, Y. Wan, T. Huang, S. Pispas, J. W. Mays, C. Wu, *Macromolecules* **30,** 7202 (1997).
7. B. Chu, *Laser Light Scattering, 2nd Ed.*; Academic Press, New York 1991.
8. W. H. Stockmayer, M. Schmidit, *Pure Appl. Chem.* **54**, 407 (1982).
9. K. Otake, H. Inomata, M. Konno, S. Saito, *J. Chem. Phys.* **91**, 1345 (1989).
10. M. Annaka, T. Tanaka, *Nature* **355**, 430 (1992).

Permanent and Reversible Gels: Transport of Guests and Guest-Host Interactions

Shulamith Schlick,[§][*] Jan Pilar[#]

[§]Department of Chemistry and Biochemistry, University of Detroit Mercy, Detroit, Michigan 48219, USA and

[#]Institute of Macromolecular Chemistry, Academy of Sciences of the Czech Republic, 162 06 Prague 6, Czech Republic

SUMMARY: We describe the application of electron spin resonance imaging (ESRI) for measuring *macroscopic* diffusion coefficients of paramagnetic guests in polymeric systems. The ESRI method is based on encoding spatial information in the ESR spectra via magnetic field gradients, and on simulating the time evolution of the spectra in order to extract the diffusion coefficient, D. This paper presents one dimensional (1D) ESRI experiments planned to study (a) the dependence of the diffusion coefficients of probes on the presence of permanent crosslinks in gels, and (b) the effect of the probe site on the transport rates in self-assembled polymers. These effects were investigated, respectively, in two types of gels: chemically crosslinked (permanent) hydrogels, and reversible gels formed by self-assembly of polymeric surfactants.

Introduction

Diffusion processes in polymer solutions and in swollen polymer networks ("gels") are of considerable importance for an understanding of polymer dynamics, and for numerous applications such as separation processes, drug delivery systems, transport across biological membranes, and prosthetic devices. The transport data reflect the interactions between polymer, solvent and guests, and the relaxation phenomena in the presence of temporary entanglements (in polymer solutions), and obstacles such as permanent crosslinks (in swollen networks). Two types of methods for measuring diffusion coefficients, D, have been described in the literature: tracer methods, which measure *macroscopic* transport on the length scale of millimeters and on the time scale of minutes or more, and *microscopic* methods, which measure the self-diffusion coefficients on the time and length scales typical for molecular motion.[1-5)]

© WILEY-VCH Verlag GmbH, D-69469 Weinheim, 2000   CCC 1022-1360/00/$ 17.50+.50/0

Imaging based on ESR, ESRI, provides information on the spatial distribution of paramagnetic diffusants and has been used for measuring diffusion coefficients.[6] In our laboratory 1D and 2D (spatial-spectral) ESRI based on nitroxide spin probes and paramagnetic Mo^V have been applied for the determination of the spatial distribution and the diffusion coefficients of paramagnetic species in ion-containing polymers, polymer solutions, crosslinked polymers swollen by solvents, and self-assembled polymers.[7-9] The method is based on encoding spatial information in the ESR spectra via magnetic field gradients, and on simulating the time evolution of the spectra in order to extract the diffusion coefficient, D. This summary illustrates the application of 1D ESRI to the measurement of transport in two types of gels: permanently crosslinked networks swollen by water ("hydrogels"), and reversible gels formed by self-assembly of polymeric surfactants. Our objectives were to study the effect of the permanent crosslinks in hydrogels on the diffusion coefficients of tracers, and the transport rates of guests in polymeric amphiphiles that self-assemble in water solutions, leading to the formation of thermoreversible gels.

Determination of the Diffusion Coefficients by 1D ESRI

ESR Imaging and Data Acquisition. The ESR imaging system in our laboratory consists of the Bruker 200D ESR spectrometer equipped with two Lewis Coils (George Associates, Berkeley, USA, type 503D), and two regulated DC power supplies (Kikusui Electronic Corp., Japan, model PAE 35-30). The coils supply a maximum linear field gradient of ≈ 320 G/cm in the direction parallel to the external magnetic field (z axis), or ≈ 250 G/cm in the vertical direction (along the long axis, x, of the microwave cavity), with a control voltage of 20 Amperes applied to each power supply. Additional details have been published.[7-9]

Simulation of 1D ESR Images. The D values for a paramagnetic guest can be determined from one pair of ESR spectra, measured respectively with and without field gradient. In the presence of a gradient G_x (G/cm) along the x axis, the ESR spectrum ("projection") $I_C(H)$ is given by the convolution integral, equation 1,

$$I_C(H) = \int_{-\infty}^{\infty} f_0(H-H') \, p(H') \, dH' \qquad (1)$$

where $f_0(H)$ is the spectrum, H is the homogeneous stationary magnetic field, $H^*=x \cdot G_x$ is the local magnetic field (at x) due to the gradient G_x, and $p(H^*)$ is the distribution of the paramagnetic centers along the direction of the gradient. The spatial distribution of the spin probe $C(x)$ can be calculated from $p(H^*)$ if the gradient is known.

In a diffusion experiment, the concentration profiles as a function time, $C(x,t)$, can be calculated from a solution of Fick's law that is appropriate for the specific sample configuration.[10] The distribution function $p(H^*)$ for a gradient G_x and at time t in the diffusion process was calculated with D and h as parameters, convoluted according to equation 1 with the ESR lineshape $f_0(H-H^*)$, and compared with the experimental lineshape; the best visual fit[8] or a least square fit[9] were used to extract the diffusion coefficient. No correction for the sensitivity profile of the ESR cavity was needed, if only the initial stages of the diffusion process were used to deduce the diffusion coefficients (in the case of a long polymer layer), or if the polymer layer was <5 mm in length.

Diffusion Coefficients in Permanent Gels

In a recent study by 2D ESRI we detected a significant difference between the diffusion coefficients of a small paramagnetic tracer (MW 186) in a permanent polystyrene network swollen by toluene and in toluene solutions containing the same concentration of linear polystyrene.[7c] This finding encouraged us to focus on the effect of permanent chemical crosslinks in gels on the diffusion coefficients of tracers.

The diffusion coefficients of various tracers were measured by 1D ESRI in lightly crosslinked poly(1-vinylpyrrolidone) gel swollen at equilibrium with water, and in the corresponding aqueous solutions of the polymer. The formulae of the tracers are shown in Chart 1.[9] For ESR measurements, aqueous solutions of linear PVP were polymerized in glass capillaries with i.d. ≈ 1 mm. The gel layer (3 to 5 mm) was topped with 0.4 μL of 10^{-2}M aqueous solution of the paramagnetic tracer, sealed by parafilm, and positioned vertically in the cavity of the ESR spectrometer, with the capillary axis parallel to the gradient direction. ESR spectra were measured at suitable time intervals in the presence and in the absence of the gradient. The effect of the gradient magnitude on the quality of the data was checked by using different gradients, 46.6 G/cm or 94.5

G/cm for most of the samples.

A typical experimental and simulated projection, together with corresponding concentration profile of the paramagnetic tracer in the sample are shown in Fig. 1. Analysis of the data indicated that the most precise data are obtained once a measurable concentration of the tracer has reached the end of the sample. In such cases the sample length can be determined with high accuracy using the positions of the high-field features in the projection, as seen in Fig. 1.

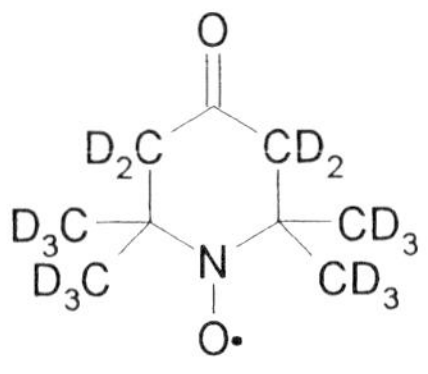

PDT (1) 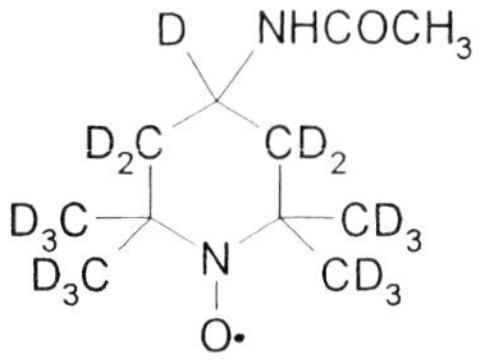   *AATEMPO (2)* 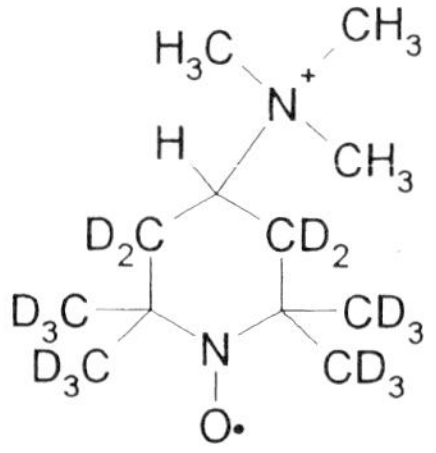   *CAT1 (3)*

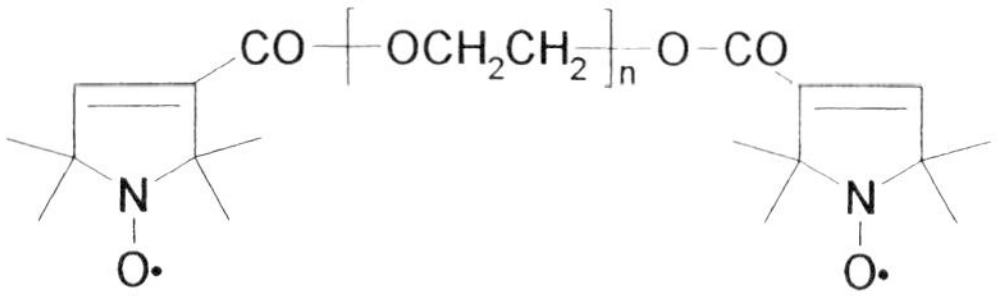

SLPEO:

(a): MW=1500 (4a)

(b): MW=1500 (4b)

Chart 1

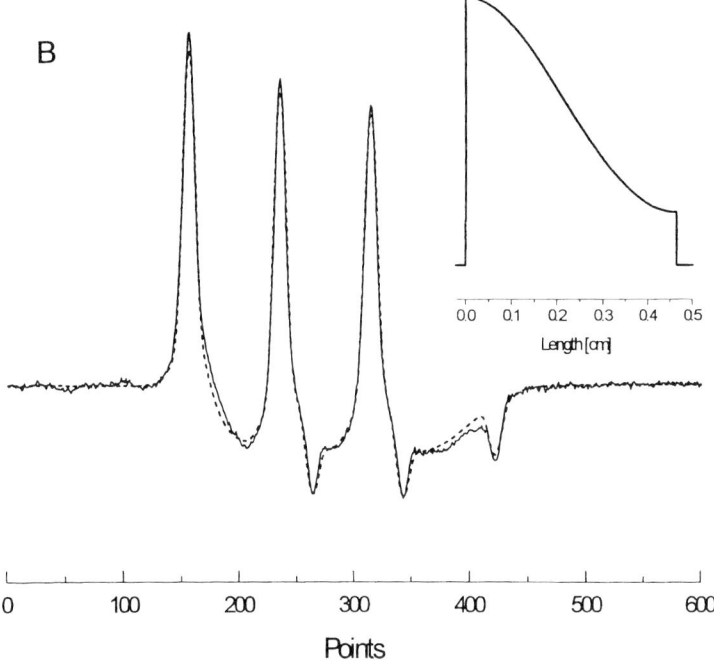

Fig. 1: Experimental (full line) ESR spectrum at time t=27,480 s after diffusion onset for tracer **4b** in the 12 % w/w PVP solution, measured with magnetic field gradient G = 46.6 G/cm, 9 scans, magnetic field sweep 120 G, and modulation amplitude 1.5 G. The simulated projection (dotted line) was calculated with D=0.83×10^{-6}cm^{2}s^{-1}, h= 0.049cm, and l=0.464 cm. The external magnetic field increases from left to right. The parameters were chosen by a nonlinear least squares fit. The inset presents the concentration profile of the tracer in the sample, with the top of the sample on the left.

The macroscopic diffusion coefficients were determined by ESRI at 300 K for the five paramagnetic tracers in the lightly crosslinked PVP gel containing in equilibrium 88 % w/w of water, in the concentrated aqueous solution of linear PVP containing the same polymer volume fraction, and in a solution of linear PVP containing 24 % w/w polymer. As expected, slower diffusion was observed for the solution containing the higher PVP concentration. No significant difference was detected between the diffusion coefficients of the four small tracers in the lightly crosslinked gel and in the solution of linear PVP containing the same polymer volume fraction (12%). A significant difference was found between diffusion coefficients in the lightly crosslinked gel and in the corresponding solution for the largest paramagnetic tracer, PEO with MW≈3000. The data clearly indicate that even a small concentrations of permanent crosslinks in the PVP gel can reduce the rate of transport of a bulky paramagnetic tracer to the value corresponding to a higher concentration (two-fold) of linear PVP in solution, where no fixed obstacles (permanent crosslinks) are present.[11]

Diffusion in Reversible Gels

The triblock copolymers poly(ethylene oxide)-b-poly(propylene oxide)-b-poly(ethylene oxide), $EO_mPO_nEO_m$, (commercial name Pluronics) have been extensively studied in the last few years, because of their interesting polymorphic structures and gel formation in aqueous solutions, and their numerous applications in drug release systems, detergents, cosmetics, treatment of burns, and water purification.[12]

We have initiated a study of the Pluronic copolymers based on the ESR spectra of spin probes that are known to intercalate in, and report on, various regions of the self-assembled system. The main objectives of these studies are to obtain *local* information on the hydration of the hydrophobic and hydrophilic domains, on the degree of order in the aggregates, and on the transport of various guests in the system. Most of our studies have focused on $EO_{13}PO_{30}EO_{13}$ (commercial name L64, Chart 2), whose phase diagram in aqueous solutions has been deduced by ^2H NMR, polarizing microscopy, and ocular inspection.[13] In the vicinity of 300 K, the main phases detected with increasing polymer content are L_1 (micellar), H (hexagonal), L_α (lamellar), and L_2 (reverse micellar). Using probes of various hydrophobicity we were able to follow changes in the hydration of the EO blocks at various distances from the hydrophobic core in the L64 aggregates on a scale of ≤2 nm, and to infer from these changes the mechanism of phase transitions.[14]

The viscosity, η, of L64 aqueous solutions at 300 K as a function of polymer content has two maxima: a strong maximum, η≈1430 P, for a polymer content of 50 % w/w (hexagonal phase), and a second maximum, η≈289 P, for a polymer content of 70 % w/w (L_α phase).[15] The viscosity of neat L64 is 7.6 P. The high viscosities are evidence for the formation of gel systems, due to the connectivity established by one chain that is part of two aggregates, for instance when each EO block in a chain is in *different* cylinders in the hexagonal phase.

We have measured the diffusion coefficients of small nitroxides PDT (Chart 1), CAT1 (the protiated analog of DCAT1, Chart 1), and 5DSE as guests in aqueous solutions of $EO_{13}PO_{30}EO_{13}$ (Pluronic L64), Chart 2. The ESR spectra have indicated that the three probes have different locations: in the water phase (CAT1), in the hydrophobic domains (5DSE)[8] and at the interface between the aggregates and the solvent (PDT).[14] The variation of the diffusion coefficients for the three probes is plotted as a function of the polymer concentration in Fig. 2.

Pluronic L64

$$HO[-CH_2CH_2O-]_{13}[-CH(CH_3)CH_2O-]_{30}[-CH_2CH_2O-]_{13}H$$
$$(EO_{13}PO_{30}EO_{13}, L64)$$

Spin Probe 5DSE

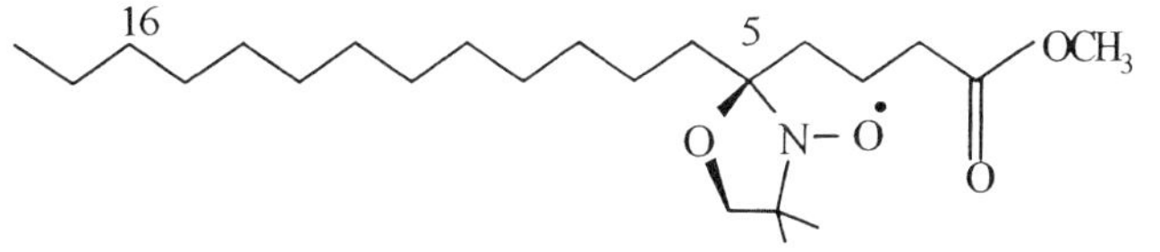

Chart 2

The dependence of D_{CAT1} on polymer content follows the expression $D=D_0\exp(-aw_2)$, where D_0 is the diffusion coefficient of the probe in neat water, and w_2 is the weight fraction of the polymer. The straight line in Fig. 2 shows the excellent agreement between this expression and the experimental results for CAT1. This behavior is consistent with the expression suggested by Phillies for the diffusion of probes:[16] $D=D_0\exp(-ac^v)$, where c is polymer concentration, and a and v are constants. Moreover, we have used a similar expression for simulating the dependence of the diffusion coefficient of pure solvents and solvent mixtures in crosslinked polyisoprene swollen by the solvents.[4] Data for CAT1 therefore suggest a behavior typical of a solvent. From these considerations it is clear that the major factor that controls the transport rate of the guests is their location: The range of D values measured in this study for tracers with similar MWs varied in the range $1.0 \cdot 10^{-5} \mathrm{cm}^2 \mathrm{s}^{-1} - 1.0 \cdot 10^{-7} \mathrm{cm}^2 \mathrm{s}^{-1}$, depending on tracer site.

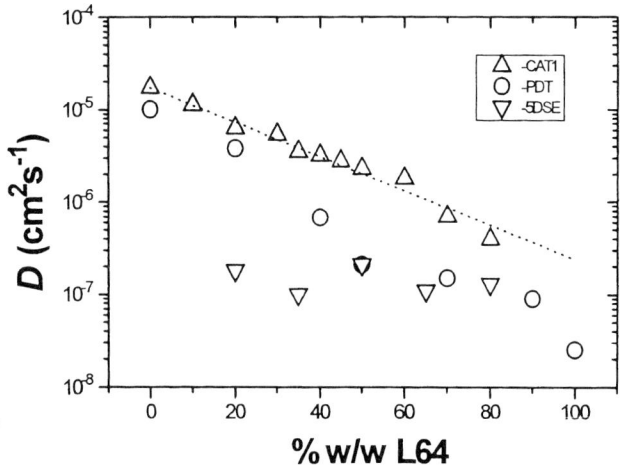

Fig.2: Variation of the diffusion coefficient, D, at 300 K for CAT1, PDT, and 5DSE with L64 concentration.

Conclusions

ESR Imaging (ESRI) is an important method for measuring *macroscopic* diffusion coefficients in hydrogels and self-assembled polymeric surfactants, and is applicable even to guests that are in low concentrations (10^{-3}-10^{-5} M), for instance drugs and other additives. The diffusion coefficients of guests in hydrogels are sensitive to the presence of permanent crosslinks (as compared to temporary entanglements): D can be reduced by a factor of ≈ 2 for large diffusants (MW 3000), even in a water-rich hydrogel (88 % w/w water). The diffusion coefficients of small guests with similar molecular weights in self-assembled polymeric surfactants can vary by two orders of magnitude, depending on the guest location.

Acknowledgments. This research was supported by the National Science Foundation (in the US) and by the Ministry of Education (in the Czech Republic).

References and Notes

1. S. Pickup, F.D. Blum, *Macromolecules* **22**, 3961 (1989)

2. B. Lindman, U. Olson, O. Söderman, in: *Dynamics of Solutions and Fluid Mixtures by NMR*, J.J. Delpuech (Ed.), J. Wiley, New York, 1995, Chapter 8, p 345

3. H. Yasunaga, I. Ando, *Polymer Gels and Networks* **1**, 83 (1993). S. Matsukawa, I. Ando, *Macromolecules* **29**, 7136 (1996) and *Macromolecules* **30**, 8310 (1997). C. Zao, S. Matsukawa, H. Kurosu, I. Ando, *Macromolecules* **31**, 3139 (1998)

4. S. Schlick, Z. Gao, S. Matsukawa, I. Ando, E. Fead, G. Rossi, *Macromolecules* **31**, 8124 (1998), and references therein

5. F.P. Duval, P. Porion, H. Van Damme, *J. Phys. Chem. B* **103**, 5730 (1999)

6. J.K. Moscicki, Y.K. Shin, J.H. Freed, in: *EPR Imaging and in Vivo EPR*; G.R. Eaton, S.S. Eaton, K. Ohno (Eds.), CRC Press, Inc., Boca Raton, 1991, Chapter 19, p 189. D. Xu, E. Hall, C.K. Ober, J.K. Moscicki, J.H. Freed, *J. Phys. Chem.* **100**, 15856 (1996)

7. (a) S. Schlick, J. Pilar, S.-C. Kweon, J. Vacik, Z. Gao, J. Labsky, *Macromolecules* **28**, 5780 (1995). (b) K. Kruczala, Z. Gao, S. Schlick, *J. Phys. Chem.* **1996**, *100*, 11427. (c) Z. Gao, J. Pilar, S. Schlick, *J. Phys. Chem.* **100**, 8430 (1996). (d) S. Schlick, P. Eagle, K. Kruczala, J. Pilar, in: *Spatially Resolved Magnetic Resonance: Methods, Materials, Medicine, Biology, Rheology, Ecology, Hardware*; P. Blümler, B. Blümich, R. Botto, E. Fukushima (Eds.), Wiley-VCH, Weinheim, 1998, Chapter 17, p 221

8. E.N. Degtyarev, S. Schlick, *Langmuir* **15**, 5040 (1999)

9. J. Pilar, J. Labsky, A. Marek, C. Konak, S. Schlick, *Macromolecules* **32**, 8230 (1999)

10. J. Crank, *The Mathematics of Diffusion*, Clarendon Press, Oxford, U.K., 1993.

11. The hydrodynamic correlation length, ξ_h, measured by dynamic light scattering (DLS) was 2.82 ± 0.05 nm in the equilibrium-swollen PVP gel containing 88 wt% of water, and 2.88 ± 0.05 nm in concentrated aqueous solution containing the same polymer volume fraction ($\Phi=0.096$ for 12 wt% of PVP in water). This result indicates that in the lightly crosslinked gel the hydrodynamic correlation length is virtually the same as in corresponding solution and that the presence of permanent crosslinks, which makes the only difference between gel and solution matrices, plays only a negligible role. Together with transport data, the DLS results suggest that the permanent entanglements in the gel are more effective in reducing the transport rates for the largest probe, compared to the entanglements present in the polymer solution

12. *Plutonic and Tetronic Surfactants*, Technical Brochure, BASF Corp., Parsipanny, NJ, 1989

13. K. Zhang, A. Khan, *Macromolecule* **28**, 3807 (1995). P. Alexandridis, D. Zhou, A. Khan, *Langmuir* **12**, 2690 (1996)

14. (a) K. Malka, S. Schlick, *Macromolecule* **30**, 456 (1997). (b) A. Caragheorgheopol, S. Schlick, *Macromolecule* **31**, 7736 (1998). (c) L. Zhou, S. Schlick, *Polymer* **41**, 4679 (2000)

15. K. Malka, S. Schlick, unpublished results

16. G.D.J. Phillies, *J. Phys. Chem.* **93**, 5029 (1989)

Radiation Curable Materials –
Principles and New Perspectives

Thomas Jaworek*, Heinz-Hilmar Bankowsky[†], Rainer Königer,

Wolfgang Reich[†], Wolfgang Schrof, Reinhold Schwalm

Polymer Research Laboratory, Marketing Coatings Raw Materials[†], BASF AG,

Carl-Bosch-Straße 38, 67169 Ludwigshafen, Germany

SUMMARY: Increasing environmental concerns and the ensuing legislation to cut emissions of volatile organic compounds (VOCs) have been major driving forces behind the development of radiation cured coatings over the past 25 years. Today radiation cured coatings are known for their good overall performance and their excellent resistance against chemical and physical surface damages. Advanced photoinitiator systems allow the light stabilisation of UV-curable formulations and the outdoor application of the coating. The rapid curing, combined with the possibility of immediate processing of the coated objects opens the way for radiation curing – in 100%-, water based-, and dual cure systems as well as for radiation curable powder coatings - for a wide variety of application.

Introduction

Increasing environmental concerns and the ensuing legislation to cut emissions of volatile organic compounds (VOCs) have been the major driving force behind the development of radiation-curing coatings over the past 25 years. Technical advances have also played an important role in the spread of this fascinating field of coatings technology. The need to cut energy and raw material costs meant that new binder types, coatings formulations, as well as application and curing techniques, had to be found.

Often success was only possible through close co-operation of scientists and engineers in allied industries. As research expanded, so did knowledge about the relationships between the chemistry of the curing process and the application properties of the cured film. Apart from the environmental and technological aspects, economics has played a deciding role in the direction radiation curing has developed.

In the early years the main applications were limited to the coating of flat substrates such as boards, paper and wood. In the past decade, however, the range of applications has expanded considerably. Today, the technology has been applied to plastics, glass and even 3D objects. Coatings for metals and exterior applications are growing in importance.

CCC 1022-1360/00/$ 17.50+.50/0

Principles of Radiation Curing

Radiation curing formulations can be grouped according to the polymerisation mechanism in radical systems and, less importantly, cationic cured systems, which will not be covered in this paper. The radical systems are based on low viscosity acrylate binders which are usually formulated solvent free and applied to substrates by various techniques. The curing is performed by exposure to high energy radiation, such as ultraviolet (UV) light or an electron beam (EB). Curing with UV radiation requires a photoinitiator in the coating formulation which starts the radical polymerisation. The main advantage of radiation curing to initiate the curing lies in the very high polymerisation rates reached under intense illumination, so that the transition from the liquid formulation to the solid coating takes place in a fraction of a second. Today UV-curing accounts for 90% of the radiation curing market, as the technical requirements for EB curing and the investment costs involved are much higher. However, apart from the photoinitiator, the raw materials used in EB- and UV-curing techniques are identical.

There are two classes of **photoinitiator** employed in UV curing, those which form radicals by hydrogen abstraction of a H-donor, e.g. as in the case with benzophenone **1**, or by homolytic cleavage of C-C-bonds, e.g. as in the case with 2,4,6-trimethylbenzoyldiphenylphosphin oxide **2** (Lucirin®-TPO).

1

2

Acylphosphine oxides have absorption bands extending up to the near UV/visible region, which make them especially suitable for the curing of thick films. The long wavelength absorption of this type of photoinitiator allows the use of light stabilisers in the UV-curable formulations, necessary for outdoor applications.

Photoinitiators are usually applied in concentration between one and 5% in the coating formulation, however, they significantly contribute to the costs of the coating formulation.

Radiation curable resins are most commonly based on acrylates as the reactive functional group. The acrylate groups are attached to oligomeric or monomeric compounds, to yield resins/binders or functional monomers/reactive thinners.

Functional monomers or reactive thinners are used to adjust the working viscosity of the coatings and the film properties, which depend on the crosslinking density, which can be controlled by the functionality of the monomers. Mono-, di-, tri and polyfunctional compounds are known. Monofunctional thinners, owing to their very strong odour, are only used in small amounts for special applications such as plastic coatings. The most important thinners contain two acrylic groups, such as hexanediol diacrylate (HDDA), dipropylene glycol diacrylate (DPGDA) and tripropylene glycol diacrylate (TPGDA). One commonly used trifunctional monomer is trimethylolpropane triacrylate (TMPTA).

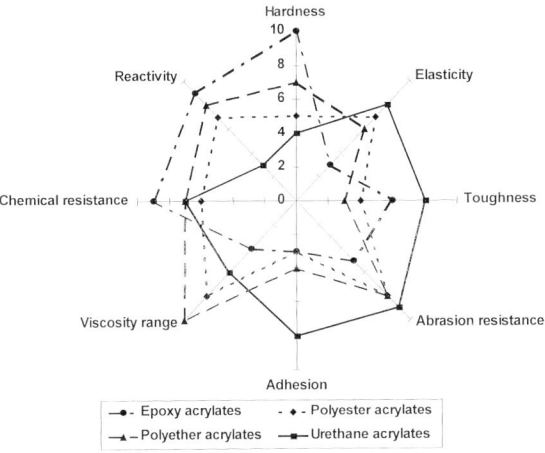

Fig. 1: Comparison of coating properties by binder classes.

Epoxy acrylates dominate the binder market, despite the fact that the viscous aromatic epoxy moieties have to be thinned with the potentially hazardous functional monomers. The more expensive aliphatic epoxy resins, on the other hand, are only used in special applications.

The wide choice of **polyester acrylates** available enables a broad range of demands to be met by the coating. These are generally low-viscous resins, requiring little or no reactive thinner.

Polyether acrylates, the least viscous group of resins, are made by esterfication of polyetherols with acrylic acid. These, too, usually do not require thinners. They can be made more reactive by tagging on amine groups.

Urethane acrylates are produced by reacting isocyanates with hydroxy alkyl acrylates, usually along with other hydroxy compounds to give the desired set of properties. Aliphatic urethane acrylates are the most expensive class of materials, however, due to their excellent weather resistance, they have become an important class of binders for outdoor applications.

Initiation	Photo-Initiation	PI	$\xrightarrow{h\nu}$	R•
	Chain-Start	R• + M	⟶	R—M•
Propagation		R—M• + M	⟶	R—M$_n$•
Transfer		R—M$_n$• + TH	⟶	R—M$_n$—H + T•
Termination	Recombination	R—M$_n$• + T/R•	⟶	R—M$_n$—T/R
	Quenching	R—M$_n$• + Q—H	⟶	R—M$_n$—H + Q
	Disproportionation	2 R—CH$_2$-CHX •	⟶	R—CH$_2$-CH$_2$X + R—CH=CHX

Fig. 2: Principle steps for a radical photopolymerisation.

During the process of curing, the photoinitiator absorbs UV light and forms radicals which add to an acrylic group of a binder or thinner to start the radical chain reaction. Within a fraction of a second thousands of acrylic groups are consumed, before the reaction stops by recombination, disproportionation or quenching. The radiation curing follows the mechanism of a radical chain polymerisation, figure 2. While monofunctional acrylic monomers give a linear polymer, higher functional binders give a polymer network.

The network density, together with the chemical structure of the polymer chains determines the properties of the coating. It is important to follow the mechanism of UV-curing, e.g. by real-time infrared (IR) spectroscopy, to learn about the reactivity of binders and to know the conversion. The degree of cure is known to correlate with the chemical resistance of the final coatings. Figure 3 depicts the network formation of a bifunctional thinner.

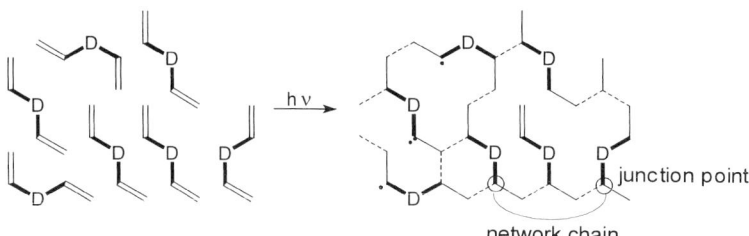

Fig. 3: Network formation with a bifunctional thinner.

While there is a good knowledge about the correlation of the mechanical properties of a polymer network with its structure, there is only a loose relationship of the polymer-network properties and the application properties of the final coating. The product properties like chemical resitstance, adhesion, hardness, toughness, elasticity, abrasion or scratch and mar resistance, influence each other and form a complex property matrix for the coatings.

New Perspectives

While the coating of wood, wood products and furniture finishing films will remain the main application area for radiation cured coatings, the coating of other substrates like plastics and metal come in focus for UV-coatings.

Radiation cured coatings on plastic substrates give a **scratch/mar resistance** comparable with high-performance organic-inorganic hybrid coatings, as shown in Figure 4. Cured under a normal atmosphere the scratch/mar resistance of pure binder systems is better than conventional 2K formulations used as automotive topcoats. Radiation curing under normal atmosphere struggles with the inhibition of the radical chain reaction by oxygen diffusing into the film. Besides using an inert atmosphere which seems inapplicable for radiation curing in some cases, surface active additives can be used to build a oxygen barrier at the uncured film surface.

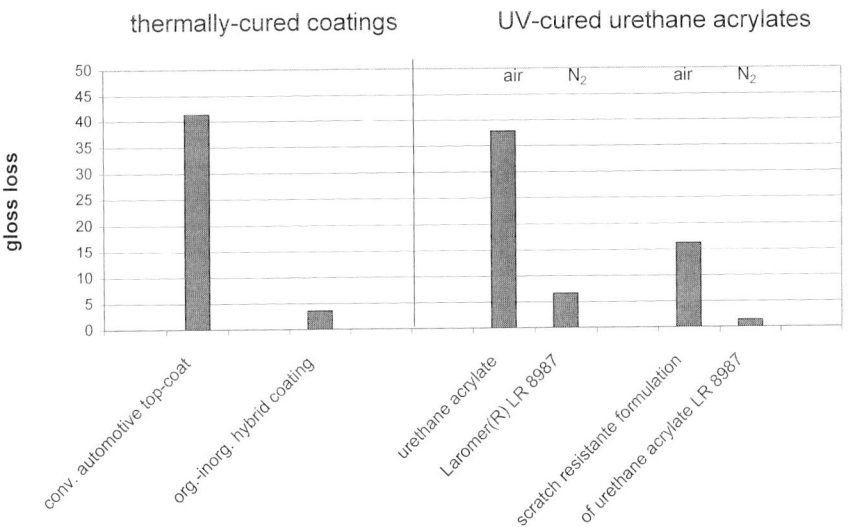

Fig. 4: Comparison of the scratch/mar resistance of coatings. The coatings have been physically damaged by a Scotch-Brite® fleece with a normal force of 84 g/cm² for 50 cycles. The scratch/resistance is inversely proportional to the gloss loss found.

The promising properties of radiation cured coatings give reasons for the reconsideration of paradigm that radiation curing can not be used for **three-dimensional objects** as the UV-light can not reach shaded regions and therefore will be left with some wet or sticky areas which will lose the protection by the coating with time. It looks as if the manufacturers of UV-equipment will offer a technical solution to this problem within the next years.

Today **dual cure systems**, combining radiation curing with a second curing mechanism are very attractive, adding excellent surface resistance against chemical and mechanical damages to the properties of a conventional coating. In most cases the areas with the highest demand on resistance can easily be reached by UV light. In principle any combination of curing mechanisms, e.g. thermal-, oxidative-, UV-cationic-curing or physical drying can be combined with UV-curing. In all cases it is necessary to balance the properties of the two networks formed. The UV curing equipment are compact and occupy relatively little space so that they may be added to an existing coating line. In comparison to thermal curing, they consume less energy and besides the environmental reasons there are economical reasons in favour of radiation curing.

Another paradigm to UV-curing is, that UV-absorbers used for light stabilisation of the final coating can not be used for UV-coatings, as the UV-absorber competes with the photoinitiator for the UV-light. This has two drawbacks for the coating, that is, that the film is not thoroughly cured and/or that the film is not stable enough for **outdoor applications**. Normal sunlight induces the photo-oxidation and degradation of the coating. UV-absorbers absorb the sunlight in the UV-region and transfer the energy to thermal relaxation rather than to radical formation. The new class of photoinitiators, the bisacylphosphine oxide have a longer wavelength absorption band and allow coating formulations for light-stabilised UV-coatings.

Based on aliphatic urethane acrylates coatings the coatings show little yellowing and a weather resistance comparable to 2K-PUR automotive coatings.

Using water as a carrier system higher molecular weight binders can be applied to a substrate in relatively low viscous formulations. The introduction of water as a 'solvent' in radiation curing contradicts the idea of a 100%-systems, and in addition lacks the advantages of a conventional aqueous coating systems introducing the more expensive UV-systems. However, higher molecular weight UV-curable binders can be used in this aqueous systems without the addition of low molecular weight thinners, which migrate into porous substrates, e.g. wood. Therefor the extractable parts of uncured monomers is significantly lower than in 100%-UV coating systems.

UV-dispersions overcome problems of classical dispersion, too, which lack chemical and surface physical resistance and show low gloss and poor blocking resistance. The higher molecular weight binders need less crosslinking during the UV-curing step to give a coating with good properties. Therefore the volume reduction during the UV-curing step is lower than with conventional systems. This results in reduced mechanical stress in the coating and a better adhesion of the coating to the substrate.

In principle it seems possible to transfer all conventional radiation curable resins into water, either as they are water soluble, or by stabilisation with an emulsifier or by addition of e.g. ionic groups allowing the resin to disperse in water. Most recently air-drying urethane acrylates dispersions were introduced which exhibit relatively good blocking resistance even before radiation curing, enabling the coated article to be handled before the final crosslinking.

The field of **radiation curable powder coatings** is still in its beginning but offers interesting perspectives. Like in the case of UV-dispersions, higher molecular weight binders can be used in the coating. Instead of the physical drying of the film to evaporate the water, the UV-curable powder is molten. The advantage to conventional thermal cured powder coatings is the separation of the melting/film-formation step and the final curing.

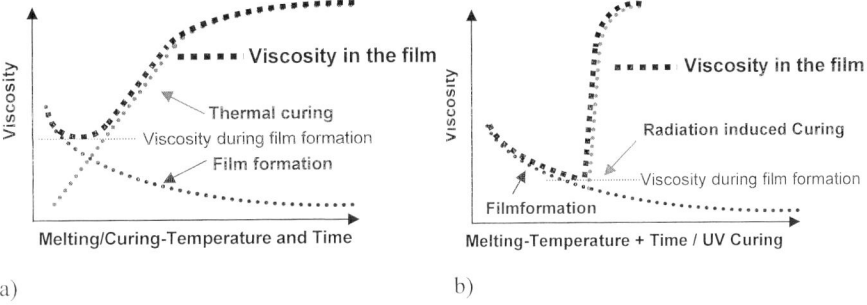

a)   b)

Figure 5: Profiles of the viscosity in the film during film formation for thermal a) and radiation b) induced curing of powder coating.

Figure 5 depicts the viscosity temperature profiles for conventional a) and radiation curable b) powder coatings. While in the first case the thermal curing starts already during the film formation, secondly, in the case of the much quicker radiation curing process the time of initiation can be choosen indepentently to the melting process of the resins. This promises advantages in terms of the levelling and the coating appearance. In combination with a fast heating IR-radiation source it seems possible to use radiation curable powder coatings for heat sensitive substrates, e.g. plastic and wood.

Outlook and Summary

Figure 6 shows the prospects of the coating markets, where radiation curable products compete with solvent based systems, with high solid, water-based and powder coatings. The relative importance of solvent based systems is declining in favour of the others, of which radiation cured products have the lowest market share. However, they promise the highest growth rate.

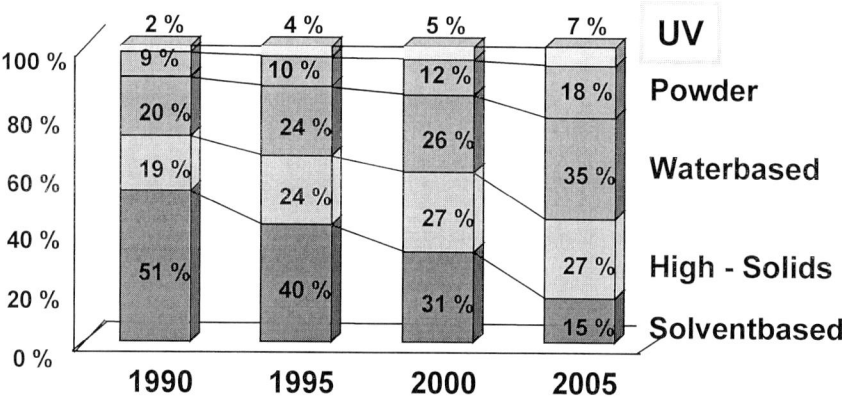

Figure 6: Market prospects for coatings as for the European paint industry. (Source: Information Research Ltd.).

In recent years, radiation curing has established itself as an environmentally friendly technology that saves both energy and material resources. In Western Europe the greatest penetration has been in the wood coatings and graphic arts sector. There is a clear trend towards low-monomer-content products that are potentially less harmful. The prospects for further expansion and introduction of radiation-curing technology into new applications look good but depends on the following becoming available commercially:

- low-viscosity, monomer-free oligomers
- low VOC raw materials
- oligomers that adhere well to critical substrates
- weathering resistant products
- improved curing equipment

The long-term success of radiation curing will require a concerned effort from all parts of the coatings industry. Problems can no longer be solved alone, but by adopting a joint approach of the raw material together with the coating formulation suppliers in close cooperation with the manufacturers of equipment for coating application and curing.

Sol-gel Transition of Biopolymer Dispersions

Katsuyoshi Nishinari

Faculty of Human Life Science, Osaka City University, Sumiyoshi, Osaka 558-8585, Japan

SUMMARY: Sol-gel transition of dispersions of biopolymers, which are used widely in food, cosmetics, biomedical and related industries, is classified by the temperature dependence of storage shear modulus. Sol-gel transition of gellan gum solution is described well by a criterion of Winter-Chambon. Too fast gelation of dispersions of konjac glucomannan in the presence of excessive amount of alkaline coagulant and/or at higher temperatures leads to a formation of gels with lower modulus. A solution of xyloglucan from which a certain amount of galactose residues is removed forms a gel on heating and reverts into a solution on further heating. The lower temperature transition of this xyloglulcan solution is induced by hydrophobic interaction.

Introduction

Gelling biopolymers have been used to modify the texture or to control the rheological properties of foods. Rheological properties of solutions of these biopolymers have been studied extensively [1,2]. Frequency and temperature dependence, and the time evolution of storage and loss shear moduli have been observed together with differential scanning calorimetry (DSC) to understand the gelation mechanism [3]. The small deformation rheology has the advantage because it is easier to obtain reproducible results and it does not break the structure being formed during gelation.

There are various biopolymers which form a gel on heating such as konjac glucomannan in the presence of alkaline coagulant or soybean glycinin in the presence of glucono-delta-lactone; solutions of both polymers form thermo-irreversible gels. Gelatin, agarose, carrageenans, gellan, and many other polysaccharides form a gel when their solutions are cooled, and these gels are thermo-reversible. Solutions of methylcellulose, xyloglucan from which some galactose residues are removed form a gel on heating, and these gels are also thermo-reversible. Recently, the interaction between different biopolymers has been studied extensively since the discovery of gel formation by mixing non-gelling agents such as xanthan and

CCC 1022-1360/00/$ 17.50+.50/0

206

galactomannan derived from locustbean. It has been expected to find other new and useful combinations to develop new gelling agents[4].

Fig. 1 shows the temperature dependence of storage modulus for four dispersions. In case (a) the gel is formed on cooling as in agarose, gellan, carrageenan, gelatin etc, and in case (b) the gel is formed on heating as in methylcellulose, methylhydroxylpropylcellulose, ovalbumin, and in case (c) the gel is formed both at lower temperatures and at higher temperatures as in the mixture of gelatin and methylcellulose, and in case (d) the gel is formed only at intermediate temperatures as in xyloglucan from which some galactose residues are removed. Every case is possible, and it suggests the infinite possibility of the science and technology of gels and gelling processes. This paper describes some recent development in understanding the sol-gel transition of three polysaccharides which show different thermal behaviours such as cases (a), (b) and (d) in Fig. 1.

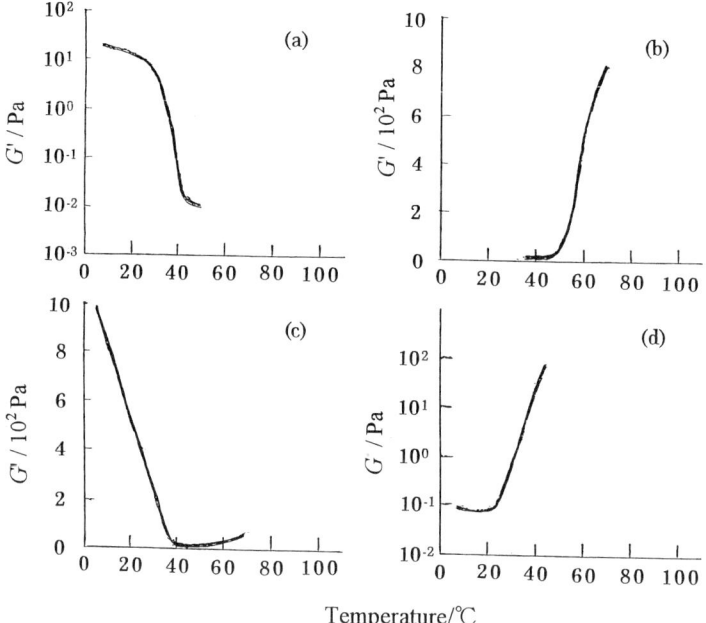

Fig. 1: Temperature dependence of storage shear modulus G' for (a) 3% solution of sodium type gellan[5], (b) 3% solution of methylcellulose[6], (c) mixed solution of 5% gelatin and 2% methylcellulose[7], (d) 1% solution of xyloglucan from which 43% galactose was removed[8]. Storage shear modulus at higher temperatures >50℃ have not been measured but should decrease with increaseing temperature.

Sol-gel transition of gellan gum solutions

Gellan gum is a microbial polysaccharide consisting of a tetrasaccharide unit, glucose-glucuronic acid-glucose-rhamnose, and has attracted much attention because it forms a transparent, heat- and acid-resistant gel.

Since even a subtle change in metal ion content or a molecular weight distribution will cause a drastic change in physico-chemical properties of polyelectrolyte solutions such as gellan gum, the common sample of sodium type gellan gum was prepared by San-Ei-gen F.F.I. and distributed to 17 laboratories with different disciplines of the working party on gellan gum organised in the Research Group on Polymer Gels affiliated to the Society of Polymer Science, Japan. This collaborative research began ten years ago using the common sample and the results were published twice[9,10]. The third common sample used was a sodium type which was easily soluble in cold water, and it was possible to determine the weight average and number average molecular weights to be $M_w = 9.5 \times 10^4$ by light scattering[11], and $M_n = 5.8 \times 10^4$ by osmotic pressure measurement[12] respectively. Results obtained by this collaborative research group by using X-ray small angle scattering, circular dichroism, NMR, ESR, DSC, dielectric and viscoelastic measurements are now in press as a special issue of the Progress in Colloid and Polymer Science. The present section concentrates mainly on rheological and DSC studies of this gellan.

The state diagram of gellan gum aqueous solution is shown in Fig. 2(a)[5]. The coil-helix transition temperature T_{ch} was determined as the temperature at which loss shear modulus G'' as a function of temperature showed a steep rise on cooling. This temperature coincided well with the transition temperature at which the molar ellipticity at 202nm changed drastically in the circular dichroism measurements[13]. The sol-gel transition temperature T_{sg} was determined as the temperature at which storage shear modulus G' took over loss shear modulus G'' on cooling, and G' showed a step-like change.

Frequency dependence of storage and loss moduli, G' and G'', for 1~3.5 wt% sodium type gellan gum aqueous solutions at various temperatures is shown in Fig. 2 (b-d)[5]. A 1% solution shows a typical dilute solution behaviour of flexible linear polymers at 0~30℃, i.e., $G'' > G'$ for all the frequencies accessible, and both moduli increase with increasing frequency (Fig. 2(b)). The frequency dependence of both moduli could be well approximated by $G'' \sim$

ω and $G' \sim \omega^2$ at low frequencies, which is a characteristic feature for dilute polymer solutions. A 2.5% solution also shows a dilute solution behaviour above $T_{ch} = 37.7°C$, but G' and G'' show a cross-over between $T_{ch} = 37.7$ and $T_{sg} = 26°C$; $G'' > G'$ at lower frequencies but $G' > G''$ at higher frequencies, a so-called concentrated solution rheological behaviour where the molecular chains disentangle during a long period of oscillation at low frequencies, and the solution behaves as a viscous liquid, whilst the molecular chains do not disentangle during a short period of oscillation at high frequencies, and their entanglement points play a role of temporary knots of three-dimensional network, and as a result the behaviour of the

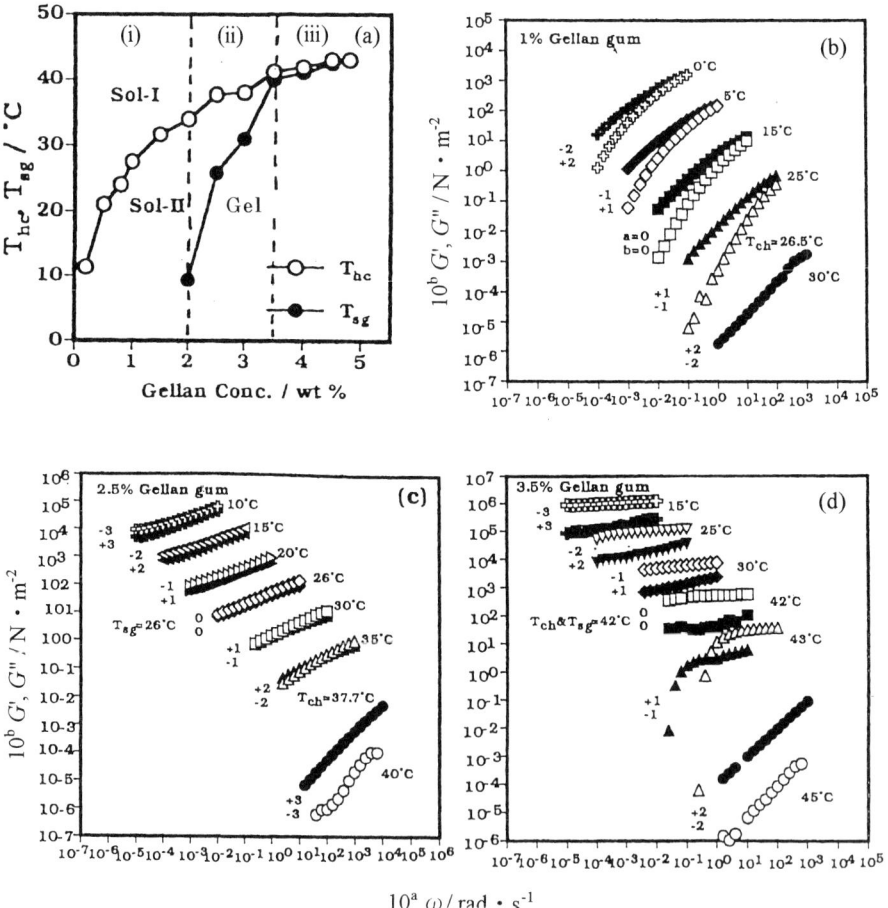

Fig. 2: State diagram (a) and frequency dependence of storage (open) and loss (closed) shear moduli for 1%(b), 2.5%(c), and 3.5%(d) gellan gum solutions. The data are shifted along both the horizontal and vertical axes by shift factors a and b, respectively to avoid the overlapping.

solution tends to that of an elastic solid (Fig.2(c)); at the sol-gel transition temperature $T_{sg} = 26°C$, the solution is in the critical gel state which Winter-and Chambon proposed, i.e., both moduli showed the same frequency dependence, $G''\sim G' \sim \omega^{0.5}$; below the sol-gel transition temperature the solution tends to show a weak gel behaviour, i.e., G' is slightly larger than G'' and both moduli are only slightly dependent on the frequency. A 3.5% solution shows a concentrated solution behaviour at $43°C$, however, it turns to an elastic gel behaviour on lowering the temperature only $1°C$ (Fig. 2(d)).

It is well known that physico-chemical properties of anionic polysaccharides such as κ-carrageenan and gellan are strongly influenced by the addition of salts[1-3]. The endothermic peak in the heating DSC curves and the exothermic peak in the cooling DSC curves shifted to higher temperatures with increasing concentration of the added salt. The difference between the exothermic peak temperature in the cooling DSC curve and the endothermic peak temperature in the heating DSC curve became larger above a certain concentration of the added salt[5]. The temperature dependence of G' was shown with a cooling DSC curve for 1% gellan gum solution in the presence of CsCl at the same cooling rate (Fig. 3). The lower temperature DSC peak in the presence of 10mM CsCl was not observed in the previous common sample because the main exothermic peak was broader in the previous sample[14]. The lower temperature DSC peak corresponded well with the sol-gel transition where G' and G'' showed a cross-over in the thermal scanning rheology. The minimum content of the salt

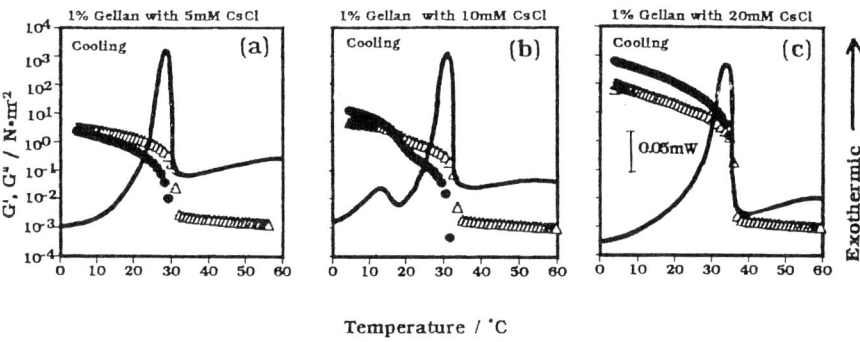

Fig.3: Cooling DSC curves and the temperature dependence of storage shear modulus G'(●) and loss shear modulus (△) at 0.1 rad/s observed at the same cooling rate (0.5°C/min) for 1 % gellan gum solutions in the presence of CsCl of various concentrations.

which induces the gelation of 1% gellan gum solution was in the following order: $Li^+ > Na^+ > K^+ > Cs^+$, which also coincides with the dynamic hydration number[5].

Effects of glucose, fructose, sucrose and trehalose on the gel-sol transition of gellan gum were also examined by rheology and DSC and all these sugars except fructose were found to promote the gel formation of gellan gum[5].

Sol-gel transition of konjac glucomannan dispersions

Konjac glucomannan (KGM), composed of mannose and glucose, and the ratio of mannose to glucose is about 1.6:1, has been used as an important ingredient in the traditional Japanese foods. Although a KGM dispersion does not form a gel without adding the alkaline coagulant, mixtures of KGM and other hydrocolloids such as κ-carrageenan or xanthan form a gel, and have been used as dessert jellies. This synergistic interaction has been studied extensively recently[15].

The frequency dependences of storage and loss shear moduli for konjac glucomannan solution from 0.35 to 1.4% show concentrated solution behaviours[16]. The time dependence of the storage and loss shear moduli at a constant temperature and at a constant frequency was observed for konjac glucomannan solutions in the presence of a constant concentration of alkaline coagulant Na_2CO_3 for KGM solutions with different molecular weights or with various polymer concentrations[17]. The increase in the storage modulus was well approximated by a first order kinetic equation $G'(t) = G'_{sat} [1 - e^{-k(t-t_0)}]$, where G'_{sat} is the plateau value of G' after a long time, k is the rate constant of gelation, and t_0 is the gelation time. The gelation time t_0 became shorter, and k and G'_{sat} became larger with increasing molecular weight or concentration. G'_{sat} varied with the concentration as $c^{2.55}$. The time dependence of the storage and loss shear moduli in the presence of a constant alkaline coagulant and at a constant frequency was observed for konjac glucomannan solutions at various temperatures from 50 to 80℃. The gelation time t_0 became shorter and the rate constant k and the saturated value of storage shear modulus G'_{sat} became larger with increasing temperature, however, G' seemed to decrease after a certain time at higher temperatures such as 75 and 80℃. Too fast gelation in the presence of excessive alkaline coagulant led to the smaller G'_{sat} although t_0 became shorter and the rate constant k became larger (Fig. 4)[17]. This may be attributed to one of the following reasons: 1) Water exudes out when the gelation proceeds too fast and the slippage occurs between the cone plate geometry and the sample dispersion, 2) Very weak ordered structure is formed when the gelation proceeds too fast and this structure is broken during the oscillatory measurement although the amplitude is very small. The similar situations have been observed in the gelation of soybean protein and κ-carrageenan.

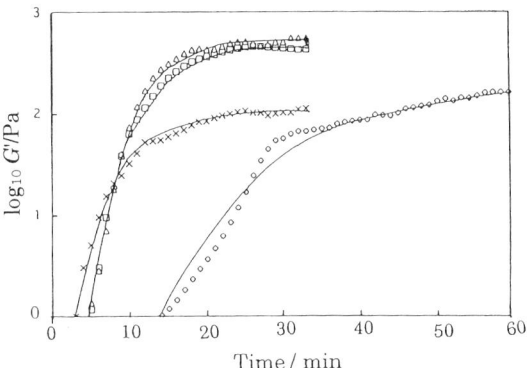

Fig. 4 : The temperature dependence of storage shear modulus of 1% konjac glucomannan dispersion in the presence of alkaline coagulant Na_2CO_3 of different concentrations. $\diamondsuit$: 0.5M, $\square$: 1.0M, $\triangle$: 1.5M, $\times$: 2.0M. Measurement temperature: 60℃, frequency: 1rad/s. Symbols represent the experimental values and the solid lines represent the calculated curves using the first order kinetic equation[17].

Sol-gel transition of xyloglucan solutions

Xyloglucan extracted from tamarind seed is a β-glucan backbone with α-D-xylosyl residues attached to the 6-position of β-glucosyl residues. It was found that when some of the galactose side chains are removed by an enzyme, the solution of xyloglucan formed a gel on heating. Shirakawa et al found that xyloglucan from which galactose residues were removed formed a gel on heating and returned into sol state on further heating, and that this transition was thermally reversible (Fig. 5)[18]. The difference between two sol states at higher temperatures and at low temperatures were not clarified yet, and should be studied in the future.

As has been observed for methylcellulose solutions[6], an endothermic peak was observed in the heating DSC curve accompanying the sol to gel transition[8]. In most other gelling polysaccharides which form a gel on cooling, an exothermic peak has been observed on cooling accompanying the sol-to-gel transition, and an endothermic peak has been observed on heating accompanying the gel-to-sol transition[3]. In these cold setting gels, the molecular forces which take part in forming junction zones are mainly hydrogen bonds while in the heat setting gels like methylcellulose or the modified xyloglucan, gel is believed to be formed by the hydrophobic interaction. The change in enthalpy accompanying the formation of hydrogen bond is negative, while the enthalpy change accompanying the formation of hydrophobic interaction can be positive or negative. The endothermic enthalpy observed for gelation of methylcellulose (16 J per one gram polymer)[19] and methylhydroxylpropylcellulose $(6.9J/g)$[20] is far smaller than that for gelatin $(24.2 J/g)$[21], κ-carrageenan $(46 J/g)$[22], and agarose $(40 J/g)$[23]. As is stated above, the latter group forms a gel on cooling by hydrogen bonds, while the former group forms a gel on heating by hydrophobic interaction. It is well known that the energy for hydrophobic interaction is about one-fourth of that for hydrogen bond[24]. The endothermic enthalpy observed for gelation of xyloglucan was $4.4 J/g$[8], and is the same order of the magnitude for the former group, suggesting that the gel of xyloglucan is formed by hydrophobic interaction.

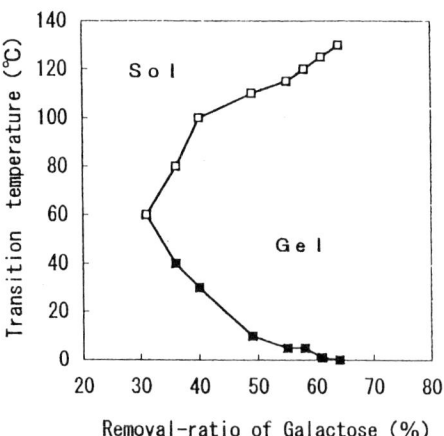

Fig.5: The sol-gel transition temperature diagram of 2% xyloglucan solution as a function of the removal ratio of galactose from xyloglucan. □: high-temperature transition point, ■: low-temperature transition point[18].

Conclusion

It has been shown that helix formation is pre-requisite for gel formation in gellan gum solution, and that gel is not formed below a certain critical concentration, and that above a certain concentration gel is formed concurrently with helix formation. This is in agreement with the classical picture proposed earlier for sol-gel transition of ι -carrageenan[25]. However, another picture for gels, a "fibrous model" is proposed recently based on atomic force microscopic obsevation[26]. It is necessary to study further using the common sample with different molecular weights with narrow molecular weight distribution. It has been shown that gel formation of konjac glucomannan dispersions on heating in the presence of alkaline coagulant proceeds faster with increasing temperature, concentration of KGM/alkaline coagulant, and molecular weight of KGM. However, too fast gelation leads to a gel with a smaller elastic modulus. This should also be clarified by further study. It has been suggested that gel formation of solutions of xyloglucan from which galactose residues are removed is induced by hydrophobic interaction. The higher temperature gel-sol transition should be studied in the future. Further study on the relation between chemical structure of biopolymers and sol-gel transition is required to develop further utilisation of these biopolymers.

References

1. A. H. Clark, S. B. Ross-Murphy, *Adv. Polym. Sci.* **83**, 57 (1987)
2. K. te Nijenhuis, *Adv. Polym. Sci.* **130**, 1 (1996)
3. K. Nishinari, *Colloid Polym. Sci.* **275**, 1093 (1997)
4. K. Nishinari (Ed.), *Hydrocolloids I: Physical Chemistry and Industrial Application of Gels, Polysaccharides, and Proteins, Hydrocolloids II: Fundamentals and Applications in Food, Biology, and Medicine*, Elsevier, Amsterdam, in press
5. E. Miyoshi, K. Nishinari, *Prog. Colloid Polym. Sci.* **114**, 68 (1999)
6. K. Nishinari, K. E. Hofmann, H. Moritaka, K. Kohyama, N. Nishinari, *Macromol. Chem. Phys.* **198**, 1217 (1997)
7. K. Nishinari, K. E. Hofmann, K. Kohyama, H. Moritaka, N. Nishinari, M. Watase, *Biorheology* **30**, 241 (1993)
8. M. Shirakawa, Y. Uno, K. Yamatoya, K. Nishinari, in: *Gums and Stabilisers for the Food Industry 9*, G.O. Phillips, P.A. Williams, D.J. Wedlock (Eds.), The Royal Society of Chemistry, Cambridge 1998, p. 94
9. Special issue on gellan, *Food Hydrocoll.* **7**, 361 (1993)

10. Special issue on gellan, *Cabohydr. Polym.* **30**, 75 (1996)

11. R. Takahashi, M. Akutu, K. Kubota, K. Nakamura, *Prog. Colloid Polym. Sci.* **114**, 1 (1999)

12. E. Ogawa, *Prog. Colloid Polym. Sci.* **114**, 8 (1999)

13. S. Matsukawa, Z. Tang, T. Watanabe, *Prog. Colloid Polym. Sci.* **114**, 15 (1999)

14. E. Miyoshi, Ph.D. Thesis, Osaka City Univ. (1996)

15. K. Nishinari, P. A. Williams, G.O. Phillips, *Food Hydrocoll.* **6**, 199 (1992)

16. M. Yoshimura, T. Takaya, K. Nishinari, *Carbohydr. Polym.* **35**, 71 (1998)

17. M. Yoshimura, K. Nishinari, *Food Hydrocoll.* **13**, 227 (1999)

18. M. Shirakawa, K. Yamotoya, K. Nishinari, *Food Hydrocoll.* **12**, 25 (1998)

19. A. Haque, E. R. Morris, *Carbohydr. Polym.* **22**, 161 (1993)

20. Y. Yuguchi, H. Urakawa, S. Kitamura, S. Ohno, K. Kajiwara, *Food Hydrocoll.* **9**, 173 (1995)

21. K. Gekko, H. Mugishima, S. Koga, *Biosci. Biotech. Biochem.* **56**, 1279 (1992)

22. M. Watase, K. Nishinari, *Macromol. Chem.* **188**, 2213 (1987)

23. M. Watase, K. Nishinari, *Macromol. Chem.* **188**, 1177 (1987)

24. J. Israelachvili (Ed.), *Intermolecular & Surface Forces, 2nd ed.*, Academic Press, London 1992

25. D. A. Rees, *Adv. Carbohydr. Chem. Biochem.* **24**, 267 (1969)

26. V. J. Morris, A. R. Kirby, A. P. Gunning, *Prog. Colloid Polym. Sci.* **114**, 102 (1999)

Macromol. Symp. 159, 215–220 (2000)

Intelligent Gel
-Surface Properties and Functions of Gels-

Jian Ping Gong, Go Kagata, Yoshihito Osada*

Graduate School of Science, Hokkaido University, Sapporo 060-0810, Japan

SUMARRY: The sliding friction of various kinds of hydrogels has been studied and it was found that the frictional behaviors of the hydrogels do not conform to Amonton's law $F = \mu W$, which well describes the friction of solids. The frictional force and its dependencies of on the load are quite different depending on the chemical structures of the gels, surface properties of the opposing substrates, and the measurement condition. The gel friction is explained in terms of interfacial interaction, either attractive or repulsive, between the polymer chain and the solid surface. According to this model, the frictional is ascribed to the viscous flow of solvent at the interface in the repulsive case. In the attractive case, the force to detach the adsorbing chain from the substrate appears as friction. Surface adhesion between glass particles and gels measured by AFM showed a good correlation with the friction, which support the repulsion-adsorption model proposed by authors.

Introduction

One of the most important researches of tribology is in the field of bio-systems, in particular, in the human and animal joints. Animal joints have friction coefficients in the range 0.001-0.03, which are remarkably low even for hydrodynamically lubricated journal bearings. However, there are several subjects to be solved. One of them is to explain why the cartilage friction of the joints is so low even in such conditions as that the pressure between the bone surfaces reaches as high as 3-18MPa and the sliding velocity is never greater than a few centimeters per second. Under such conditions, the lubricating liquid layer can not be sustained between two solid surfaces and the common hydrodynamic lubrication does not work.

Animal cartilage consists of a three-dimensional collagen network filled with a synovial fluid. We consider that the role of solvated polymer network existing in extracellular matrix of the cartilage cell as a gel state is critically important in the specific frictional behavior of cartilage. In order to investigate the general features in tribology of solvated polymer matrix, friction of various kinds of hydrogels were investigated. The gel frictions on solid surfaces have been attempted to explain in terms of the adhesion-repulsion model together with the experimental results by AFM.

Experimental Results

          CCC 1022-1360/00/$ 17.50+.50/0

Fig.1a shows the frictional force F in air for the non-ionic synthetic gel, poly(vinyl alcohol) (PVA); partially charged polysaccharide gel, gellan; fully charged polyelectrolyte gel, sodium salt of poly(2-acrylamido-2-methylpropanesulfonic acid) (PNaAMPS) on the glass plate. The friction coefficient, $\mu = F/W$, is calculated and shown in Fig.1b. F is almost constant (gellan gel), slightly increases (PVA gel) and strongly dependent (PNaAMPS) with the load over the observed range of load.

The friction coefficient μ for these gels accordingly shows unique load dependencies, which is quite different from those of solids. μ of PNaAMPS gel is constant over the change of the load, similar to those of rubber, but the value of μ is as low as 0.002, which is two orders of magnitude lower than those of solids. μ of PVA and gellan gels decreases with the increase of load. These results demonstrate that the gel friction does not simply obey Amonton's law of $F = \mu W$, in which μ is a material constant, and shows that the chemical structures of gels have a strong effect on the friction behavior.

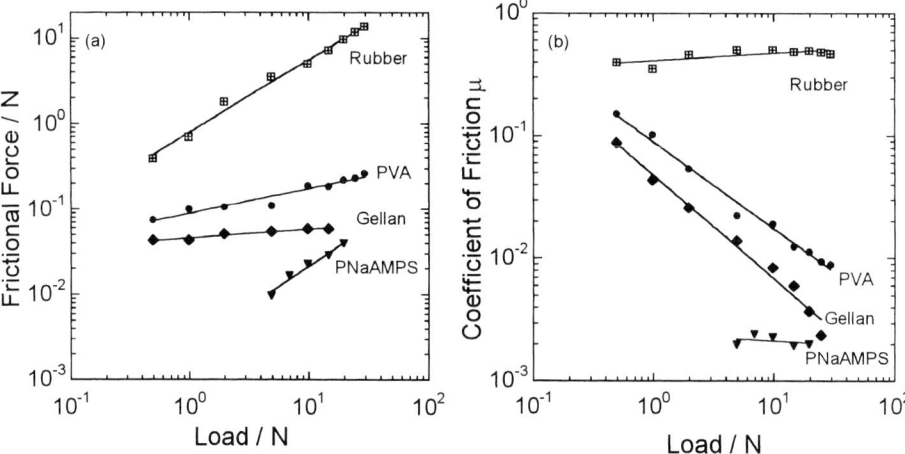

Fig.1 Dependencies of friction(a) and the coefficient of friction(b) on load for various kinds of hydrogels slide on glass surface in air. Sliding velocity: 7mm/min.; Sample contact area: PVA, gellan, and rubber, 3x3cm², PNaAMPS, 2x2cm². Degree of swelling q: PVA, 17; gellan, 33, PNaAMPS, 15.

The gel showed a much lower friction in water. Fig.2 shows the results in water measured at a velocity of 90mm/min, which is more than one order of magnitude larger than that in Fig.1 in order to obtain a friction force large enough to be measurable. The friction of PVA gel showed slightly stronger load dependence than that in air. The frictional force of poly(2-acrylamido-2-methylpropanesulfonic acid) (PHAMPS) gel is much lower with two load dependence profiles: it shows strong load dependence at low load, and less strong load dependence at high load that is similar to behavior of PVA. Such a load dependence of friction observed by PHAMPS gel was reversible.

However, if PHAMPS is allowed to slide on a Teflon plate, the friction-load profile remarkably changes. As shown in fig.2, the behavior of PHAMPS on Teflon is similar to that of PVA on glass showing a monotonous increase with the load but with a much higher value. These data demonstrate that besides the chemical structure of the gel, the surface property of the opposing substrate are crucial in friction and the interaction of the gel network with the substrate should be taken into consideration.

Repulsion-Adsorption Model

The gel friction can be explained in terms of polymer-solid surface repulsion and adsorption. When a polymer gel is placed in contact with a solid wall, the polymer chain would be either repelled from or adsorbed on the solid wall depending on the strength of the interaction between them in relative to the solvent. In the former case, a solvent layer is formed at the interface and the viscous flow of this solvent layer will make a dominant contribution to the frictional force. In the latter case, however, the adsorbing chain will be stretched when the solid surface is

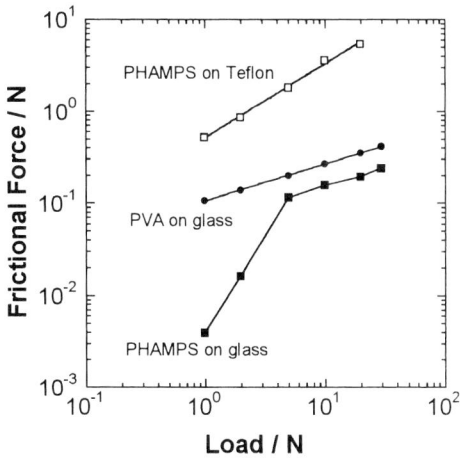

Fig.2 Load dependence of gel friction in water. Sliding velocity: 90mm/min., gel size: PVA, 3x3x0.8cm³; PHAMPS, 3x3x0.6 cm³. Degree of swelling: PVA, q=10; PHAMPS, q=15.

in motion relative to the gel. The elastic force increases with the deformation and detaches the adsorbing polymer chain from the substrate, thus it will in turn appear as the frictional force. According to this consideration, we have tried to deduce the friction theoretically.

1) Repulsive substrate

The viscous flow of the solvent layer at the interface should obey Newton's law and that of solvent in the polymer network can be expressed by the Debye-Brinkman equation[3] where the effect of polymer network is represented by a distributed force. By solving the Debye-Brinkman equation together with the Newton equation, the hydrodynamic lubrication between a repulsive solid surface and a gel is obtained[4]

$$f = \frac{\eta v}{\xi_g + \sqrt{K_{gel}}} \tag{1}$$

where η is the viscosity of solvent, v is the sliding velocity, ξ_g is the solvent layer thickness or the repelled distance of the gel from the solid surface against the normal pressure P, K_{gel} is the permeability of the gel. This result indicates that the equivalent non-slippery boundary of a gel surface is located at a depth of $\sqrt{K_{gel}}$ from the surface.

ξ_g can be determined from the interfacial free energy. The work done by the solid surface to repel the gel from the surface against the normal pressure should be equal to the increase in the interfacial free energy A, that is,

$$A - A_0 \cong P\xi_g \tag{2}$$

where A_0 is the interface energy between the substrate and the pure solvent. Supposing that the interfacial free energy between the gel and the substrate is the same as that between a polymer solution with a corresponding concentration and the substrate,

$$A - A_0 \cong \Pi_0 \xi \tag{3}$$

from the scaling theory[5]. Here Π_0 is the osmotic pressure and ξ is the correlation length of the polymer solution. On the other hand, K_{gel} can be related to ξ as $\xi \cong \sqrt{K_{gel}}$. From the above results, the friction for the repulsive substrate is

$$f \cong \eta v (E/T)^{1/3} \frac{(P/E)}{1 + (P/E)/(1 + P/E)^{1/3}} \tag{4}$$

in consideration of the deformation of the gel under the pressure P. Here E is the elastic modulus of the gel.

2) Attractive substrate

The friction mechanism in this case is quite similar to what occurred when rubber slides against a hard substrate. That is, the friction is due to the polymer chain detachment. We can follow the outline of the model proposed by A. Schallamach for the case of dynamic rubber friction. The dynamic friction per unit area due to adsorption can be expressed as[7]

$$f \cong \frac{Tv\tau_f}{R_F^4} \frac{(\tau_b/\tau_f)^2}{(\tau_b/\tau_f + 1)} \tag{5}$$

where τ_b and τ_f are the average life-times of the polymer chain in the adsorbing state and in the free state, respectively. R_F is the radius of the polymer chain. The adsorption and desorption are attributed to thermal fluctuation. The stretching of the chain would favor desorption, that is, it would lower the energy difference between the adsorbing state and the free state by a value of F_{el}, which is the elastic energy of one stretched chain. So the transition rate from the adsorbing state to the free state becomes

$$\rho = \tau_f^{-1} \exp[-(F_{ads} - F_{el})/T] \tag{6}$$

The above relation suggests that the adsorption time τ_b is a function of the deformation rate, or the sliding velocity. Under the normal pressure P, the adsorption energy would increase and it is found

$$F_{ads} \cong \begin{cases} T\phi^{-1/2}\delta(1+P/E)^{1/3} & for \quad \delta \ll \phi^{1/2} \\ T\phi^{-5/4}\delta^{5/2} & for \quad \delta \gg \phi^{1/2} \end{cases} \tag{7}$$

here, $T\delta$ is the effective attraction energy between a monomer unit and the surface in respect to the solvent and ϕ is the volume fraction of the polymer. Different with that of a rubber friction, the hydrodynamic component also exists in this case. The total friction for the attractive substrate of weak adsorption at small pressures is

$$f = \eta v(E/T)^{1/3}[\frac{1}{2} + (1+\frac{3}{2}\delta\phi^{-1/2})(1+P/E)^{1/3}] \tag{8}$$

Discussion

Fig.3 shows the calculated curves of the relations between the friction and the normal pressure for a gel sliding on solids. The theoretical analysis showed that when the interaction between a gel and a substrate is repulsive, the frictional force is small with a linear dependence on the load. In the case of the attraction, the frictional force is high with weak load dependence if the attraction is not very strong. With the increase in the attraction, the friction increases and the load dependence becomes stronger. Comparing the observed frictional force in Fig.1 and Fig.2 with the theoretical prediction, we have found that the frictional forces of PNaAMPS gel and PHAMPS gel sliding on a glass substrate

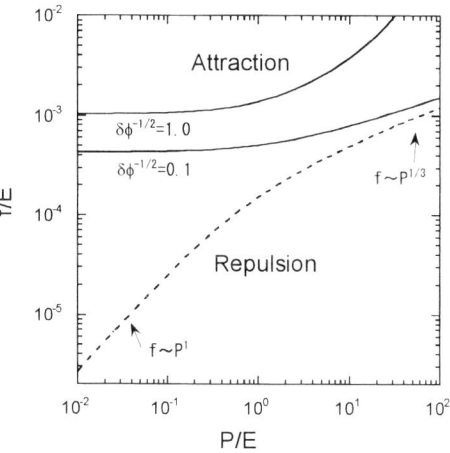

Fig.3 Pressure dependence of the frictional forces for gels sliding on a repulsive substrate (dotted line) as well as on attractive substrates with various $\delta\phi^{-1/2}$ (solid lines). $v=10^{-4}$m/s, $\phi=0.04$, a=3x10^{-10}m, E=10^5Pa, $\eta=10^{-3}$Ns/m^2, T=4.14x10^{-21}J (300K).

is quite low with strong pressure dependence. This suggests that the interaction between the PNaAMPS or PHAMPS gel with the glass surface is repulsive. The load dependence and the higher friction values of PVA gel on glass substrate and the PHAMPS gel on Teflon substrate suggest the attractive interaction.

In order to confirm the above prediction, the adhesive forces of these gels with the glass surface have been measured using the colloidal probe AFM. We found that PVA gel exhibits a large adhesive force in out-ward measurement examining the adhesive force, and gellan gel has a weak repulsion while PHAMPS gel and PNaAMPS gel show a strong electrostatic repulsion with the substrate. The repulsions are apparently due to the interaction between anions on the gel network and on the glass surface since it has negative Zeta potential in pure water. The weak repulsion between gellan and the glass could be attributed to the low charge density of $-COO^-$ in the gellan gel.

The results of AFM measurement well coincides with that of the frictional force which increases in an order of PNaAMPS<gellan<PVA. That is, the stronger the repulsion, the lower the friction. These results are in agreement with the modelling in which the predominant effect of the polymer-substrate interaction was postulated.

References

1) J. P. Gong, M. Higa, Y. Iwasaki, Y, Katsuyama, Y. Osada, *J. Phys. Chem.* B, **101**, 5487 (1997)
2) J. P. Gong, Y. Osada, *J. Chem. Phys.*, **109**, 8062 (1998)
3) H. C. Brinkman, *Physica*, **13**, 447 (1947)
4) J. P. Gong, G. Kagata, Y. Osada, *J. Phys. Chem. B*, **103**, 6007 (1999)
5) P. G. de Gennes, *Scaling Concept in Polymer Physics*, Cornell University Press, Ithaca, N. Y. (1979)
6) A. Shallamach, *Wear*, **6**, 375 (1963)
7) J. P. Gong, Y. Iwasaki, Y. Osada, K. Kurihara, Y. Hamai, *J. Phys. Chem. B*, **103**, 6001 (1999)

Development of Polymer-Based Lithium Secondary Battery

Noboru Oyama

Department of Applied Chemistry, Faculty of Technology, Tokyo University of Agriculture and Technology, Koganei, Tokyo 184-8588, Japan

SUMMARY: Novel polymer composites based on polydisulfide compounds are developed as a high energy density cathode material for lithium rechargeable batteries. A polymer composite composed of 2,5-dimercapto-1,3,4-thiadiazole (DMcT) and conducting polymer polyaniline (PAn) on a copper current collector provides high charge density exceeding 225Ah/kg-cathode with average discharge voltage at 3.4V. The composite cathode showed excellent rate capability and cyclability (>500 cycles). Surface analysis and electrochemical studies indicate that a DMcT-Cu complex plays an important role in the observed improvement of the battery performances with a copper current collector. Large increase in the charge density to 550Ah/kg-cathode is achieved by adding elemental sulfur (S_8) to the DMcT/PAn composite cathode.

Introduction

The last decade has witnessed rapid but steady spread of various portable electronic devices with reducing the size and weight. Because those advanced devices such as cellular phones almost exclusively depend on rechargeable batteries for power source, increase in energy storage capability of batteries has been pursued to downsize the devices with extending operation time. The strong demand for a compact, lightweight rechargeable battery with high energy density has promoted the development of novel electrode materials : for example, lithium cobalt oxide and carbon intercalating materials for lithium ion batteries. Recently, in addition to inorganic compounds, a functional polymer material is expected as one of promising candidates for the electrode materials with high energy density, as well as other major components of rechargeable batteries such as electrolytes and separators as shown in Figure 1. It is expected that the use of various polymer materials instead of inorganic compounds will make rechargeable batteries thinner and lighter in weight and also less hazardous to environment.

222

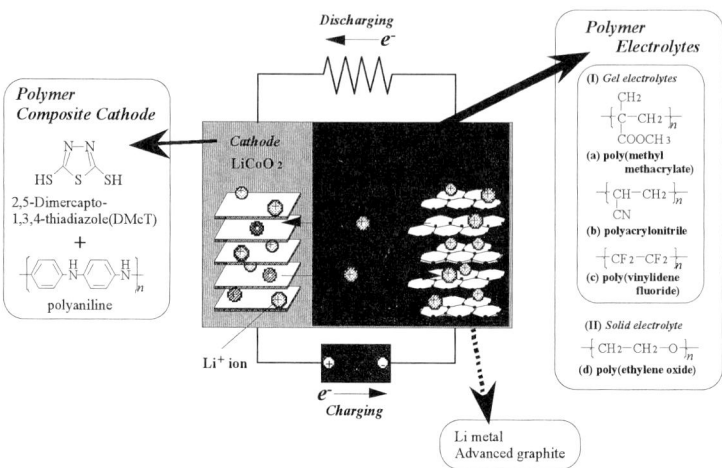

Fig. 1: Schematic depiction of the charge-discharge processes of a conventional lithium-ion battery and the structure of representative polymer materials for a polymer-based lithium battery.

Polydisulfide, which is an oxidation product of an organosulfur compound with multiple thiol groups (-SH) such as 2,5-dimercapto-1,3,4-thiadiazole (DMcT, Fig. 1), is one of promising candidates as a lightweight cathode material with high charge capacity. A polydisulfide compound is capable of storing two charges every disulfide bond (-S-S-) based on the reversible transformations between a disulfide bond and thiolates (or thiols). Based on the reaction, theoretical charge capacity of poly(DMcT) can be calculated to be 362Ah/kg-DMcT (330Ah/kg, including two lithium ions as a counterion for the monomers).

We have conducted experimental and theoretical studies on composite polymer materials composed of polydisulfide and conducting polymer as a cathode material for lithium rechargeable batteries[1-8]. In this paper, we will present our recent results obtained for a polymer composite cathode composed of DMcT and conducting polymer polyaniline (PAn) which accelerates sluggish redox reactions of DMcT at room temperature. Improved battery characteristics of the DMcT/PAn composites obtained when using a copper substrate as a current collector for the composite cathode are presented. Preparation and properties, especially high energy density, of a composite cathode material containing elemental sulfur (S_8) are also described.

DMcT/PAn Polymer Composite Cathode with Copper Current Collector

A DMcT molecule can store two charges based on the oxidation (charging) and reduction (discharging) reactions which are accompanied by reversible transformations between the thiol and disulfide as shown in Scheme 1:

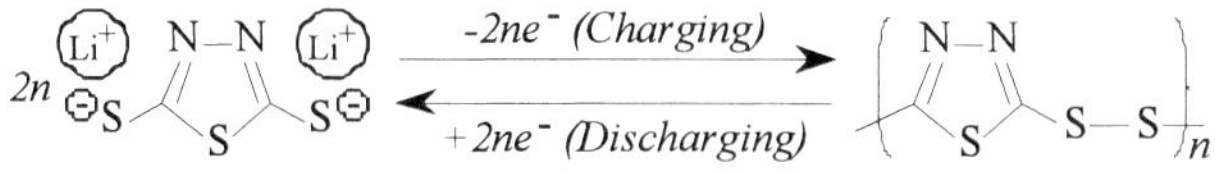

Scheme 1 Oxidation (charging) and reduction (discharging) reactions of DMcT and accompanying formation and break of the disulfide bond.

We have already reported that PAn accelerates sluggish kinetics of the DMcT redox reactions by virtue of electrocatalytic activity of PAn and making a well-mixed composite of DMcT and PAn, a cathode material with high energy density is obtained[1]. Composite formation with PAn solves another problem with organosulfur compounds as electrode active materials, i.e., PAn imparts electron conductivity to the composite cathode. PAn also serves as an electrode active material (100Ah/kg-PAn with BF_4^- as a dopant). Charge capacity of a DMcT/PAn composite was experimentally estimated to be 185Ah/kg-cathode, which corresponds to 80% of the theoretical capacity of 224Ah/kg-cathode. Average discharge voltage was 3.4V against a Li metal anode[1].

Although charge-discharge characteristics (e.g., cyclability) of DMcT were significantly improved by using as a composite with PAn[1], charging-discharging rates remained too low to use the DMcT/PAn composite as practical cathode materials. However, addition of a polypyrrole derivative resulted in the increase in the charging rate[9]. Polypyrrole derivatives are expected to serve as a subsidiary molecular current collector as well as electrode active material.

We have recently found that use of a copper substrate as a current collector for DMcT/PAn composite cathode improves rate capability for both discharging and charging, compared to other current collectors such as titanium, gold and carbon[10,11]. Furthermore, when a copper current collector was used, apparent charge capacity of the composite cathode was estimated to be 233Ah/kg-cathode, exceeding theoretical capacity estimated based on the weights of DMcT and PAn in the composites[10]. The apparent increase in the charge capacity suggested that not only DMcT and PAn but also the copper serve as a cathode active material and contribute to the observed capacity[10,11].

Recently, surface analysis has shown that DMcT strongly interacts with copper[12,13]. Surface analysis using phase measurement interferometric microscopy (PMIM) and quartz crystal microbalance (QCM) showed that copper surface spontaneously dissolved in a solution containing DMcT[13]. Using electrochemical QCM, X-ray photoelectron spectroscopy (XPS) and IR-Raman spectroscopy, it was found that copper was mainly oxidized to cuprous ion (Cu^I)[12]. It was also observed that surface of a copper current collector was covered with a thin layer containing DMcT and copper ions from XPS. Cyclic voltammetric studies have confirmed that $DMcT_2$-Cu complex showed electrocatalytic activity toward the redox reactions of DMcT[14].

Based on the results described above, it is expected that DMcT-Cu complexes accelerate the electron-transfer reactions at the interface between the composite and current collector as well as within the composite as shown in Scheme 2.

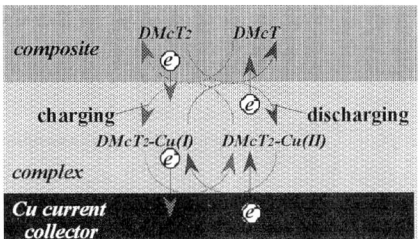

Scheme 2 A proposed mechanism for the acceleration of the DMcT redox reactions by DMcT-Cu complex formed on a copper current collector.

DMcT/PAn Composite Cathode Containing Elemental Sulfur

As shown above, the composite cathode composed of DMcT and PAn with a copper current collector possess much higher charge capacity than $LiCoO_2$ with good rate capability and cyclability. In this study, to further increase the energy density of the DMcT/PAn composite cathodes, elemental sulfur (S_8) was added to the composites.

S_8-containing composites were prepared by adding S_8 to inky solutions dissolving DMcT and PAn at high concentration. The obtained viscous solution containing S_8 was applied to current collector and dried under vacuum. A test cell was assembled using the S_8-containing composite ($DMc/PAn/S_8$) as a cathode, a foil of lithium metal as an anode and polyacrylonitrile gel containing 0.1M $LiBF_4$/propylene carbonate and ethylene carbonate as an electrolyte.

The theoretical capacity of S_8 is 1675Ah/kg-S_8, assuming that each sulfur atom is reduced to S^{2-} (16 electrons per S_8). However, in general, S_8 is reduced to a radical anion, $S_3^- \cdot$, at about +2.6V vs. Li/Li^+. $S_3^- \cdot$ is further reduced at more negative potentials, but it is mostly impossible to electrochemically reduce to $8S^{2-}$, i.e., 2-electron reduction of every sulfur atom of S_8[15]. DMcT/PAn/S_8 composite cathodes prepared in this study, on the other hand, showed high discharge capacity exceeding 550Ah/kg-cathode with average voltage at 3.5V as shown in Fig. 2[16]. The value of the discharge capacity implied that the sulfur

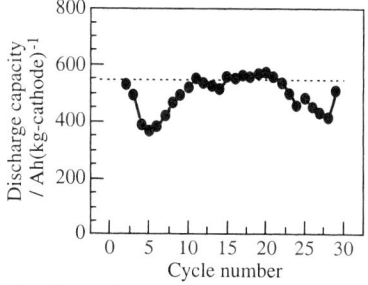

Fig. 2: Plot of discharge capacity vs. cycle number of a representative test cell with S_8 containing DMcT/PAn composite cathode.

226

was reduced to S⁻ state in the composite. The observed unusual electrochemistry of S_8 in the DMcT/PAn/S_8 composite is possibly due to strong interactions among the components of the composite.

Conclusion

Organic composites based on thiol-compounds showed a high promise as lightweight cathode materials for lithium batteries. Development of these polymer cathodes will contribute to increase energy density of lithium batteries. Furthermore, using a polymer gel electrolyte, lithium batteries will be fabricated to be thinner and unit cells can be stacked. Lithium rechargeable batteries with polymer-based cathodes and gel electrolytes will be useful as power sources for not only portable electronic devices but also for electric vehicles and space crafts.

Acknowledgement

The authors are grateful to Drs. T. Tatsuma, F. Matsumoto, J. M. Pope, Q. Chi, E. Shouji, S. C. Paulson and T. Sotomura. This study was supported in part by the Proposed-Based New Industry Creative Type Technology R&D Promotion Program from the New Energy and Industrial Technology Development Organization (NEDO) of Japan.

References

1. N. Oyama, T. Tatsuma, T. Sato, T. Sotomura, *Nature* **373**, 598 (1995)
2. A. Kaminaga, T. Tatsuma, T. Sotomura, N. Oyama, *J. Electrochem. Soc.* **142**, L47 (1995)
3. E. Shouji, N. Oyama, *J. Electroanal. Chem.* **410**, 229 (1996)
4. E. Shouji, H. Matsui, N. Oyama, *J. Electroanal. Chem.*, **417**, 17 (1996)
5. T. Tatsuma, H. Matsui, E. Shouji, N. Oyama, *J. Phys. Chem.* **100**, 14016 (1996)
6. E. Shouji, Y. Yokoyama, J. M. Pope, N. Oyama, D. A. Buttry, *J. Phys. Chem.*, B **101**, 2861 (1997)
7. T. Tatsuma, Y. Yokoyama, D. A. Buttry, N. Oyama, *J. Phys. Chem.*, B **101**, 7556 (1997)
8. N. Oyama, T. Tatsuma, T. Sotomura, *J. Power Sources* **68**, 135 (1997)
9. T. Tatsuma, T. Sotomura, T. Sato, D. A. Buttry, N. Oyama, *J. Electrochem. Soc.* **142**,

L182 (1995)

10. T. Sotomura, T. Tatsuma, N. Oyama, *J. Electrochem. Soc.* **143**, 3152 (1996)

11. N. Oyama, J. M. Pope, T. Sotomura, *J. Electrochem. Soc.* **144**. L47 (1997)

12. Q. Chi, T. Tatsuma, M. Ozaki, T. Sotomura, N. Oyama, *J. Electrochem. Soc.* **145**, 2370
(1998)

13. F. Matsumoto, M. Ozaki, Y. Inatomi, S. C. Paulson, N. Oyama, *Langmuir* **15**, 857 (1999)

14. F. Matsumoto, Y. Inatomi, N. Oyama, in preparation (1999)

15. J. Paris and V. Plichon, *Electrochim. Acta*, **26**, 1823 (1981)

16. N. Oyama, S. Hirakawa, O. Hatozaki, T. Sotomura, in preparation (1999)

Macromol. Symp. *159*, 229–245 (2000)

Polymer Battery R&D in the U.S.

Ralph J. Brodd[a], Weiwei Huang [b] and James R. Akridge [b]

(a)Broddarp of Nevada, Inc., 2161 Fountain Springs Dr., Henderson, NV 89014, U S A

(b) Eveready Battery Company, Inc., 25225 Detroit Rd., Westlake, Ohio 44145, U S A

SUMMARY: Polymer electrolytes that have been developed for battery applications fall into two general classes, neat or "pure" polymer and plasticized or gel in which the polymer is combined with a conducting organic electrolyte. The polyethylene oxide (PEO) and its modifications are typical of the "pure" polymer electrolytes. They have poor conductivity at room temperatures, but at elevated temperatures, their conductivity is of the order of 10^{-3} to 10^{-4} S/cm. The PEO electrolytes have found application in the high temperature (>60°C) lithium metal anode battery systems. The high temperature necessary for good operation makes them unsuitable for use in small consumer appliances. The polymer electrolyte battery development activities have resulted in several high performance battery systems now just entering the market. Not all of the developments have resulted in commercial cell production. The commercialization activities of high performance lithium-ion (Li-Ion) batteries have been based on two general plastic polymer systems: poly-vinylidene difluoride-hexafluoropropylene copolymer (PVdF-HFP) and polyacrylates. The polymer cells are expected to have advantages in manufacturing, flexibility, thin cell formats and lightweight packaging. Important parameters in PVdF gel electrolyte performance include the electrolyte type (combination of organic carbonates), temperature, and HFP copolymer content. Li-Ion coin cells fabricated with a polyolefin separator with either liquid electrolyte or with the PVdF gel polymer electrolyte have equivalent performance.

Introduction

Polymer batteries have been under intense research and development over the past 10 years. This year several commercial Li-Ion polymer batteries were introduced to the market place. Batteries have a set of general criteria for commercialization as listed in Table I. Polymer battery systems are capable of meeting all of these general criteria.

Polymer batteries are lighter weight than the equivalent non-polymer battery because they utilize a low cost, light weight polymer-foil container to replace the thick walled steel or aluminum can of non-polymer cells. Polymer batteries offer shape flexibility in design and are capable of very thin, large footprint constructions. Polymer cells are internally bonded and do not require external pressure on the electrode separator interface for good operation. They are essentially insensitive to shock and vibration.

   CCC 1022-1360/00/$ 17.50+.50/0

Table I. General Criteria for Commercial Batteries

Attribute	Benefit
High Energy Density	Smaller More Powerful Devices
Light Weight	Portable, Hand Held Devices
Long Life	Lower Life Cycle Costs
Flexibility	Adaptable for Many Applications
Safe Operation	Reduced Commercial Risk
Environmentally Friendly	Easy, Low Cost Disposal

Polymer electrolytes for use in batteries must be resistant to oxidation and reduction by strong oxidizing agents such as lithium cobalt oxide and by strong reducing agents such as lithium metal. This means that the range of stability is at least 4.5 volts vs. lithium. In addition, this resistance to oxidation and reduction applies over the temperature range of –40C to +85C. It is preferred that the polymers have good mechanical strength for ease in manufacturing. They can form conductive electrolytes either neat or when plasticized by addition of organic electrolytes.

Polymer Electrolyte Development

The phenomenon of complexation of salts with polyethylene oxide (PEO) or polypropylene oxide (PPO) was observed in the mid-1960s[1]. Almost ten years later, P.V. Wright and coworkers discovered that PEO and PPO complexed with alkali metal salts exhibit high ionic conductivity at elevated temperatures[2,3]. For example, the ionic conductivity of a $(PEO)_{4.5}KSCN$ complex reached roughly 10^{-3} S/cm at $120°C$. Shortly afterward, M. B. Armand and coworkers recognized the potential of these polymer electrolytes for use as separators in solid state batteries, especially in rechargeable lithium batteries[4]. Since that time, considerable effort has been directed towards the development of polymer electrolytes for use in lithium cells[5-8]. Although the ionic conductivity of the archetypal PEO solid polymer electrolytes (SPEs) is high at elevated temperatures, their conductivity is low (10^{-6} to 10^{-8} S/cm) at ambient temperatures[2]. A room temperature conductivity of at least of 10^{-3} S/cm is needed for a polymer electrolytes to function well in consumer battery systems.

Approaches to enhance the room temperature conductivity of the archetypal PEO-based "dry" SPEs include: reduction of the crystallinity level of the polymer electrolytes[9], use of polymers that have flexible backbones[10, 11], attachment of flexible pendent polar groups to the main polymer chains[12], use of special salts with large anions to form more dissociated species in polymer electrolytes[13], preparation of ionic rubbers (polymer-in-salt)[14], addition of inorganic fillers[15] etc. These approaches have not resulted in "dry" (no added liquids) and easily processible polymer electrolytes with good room temperature conductivity.

The incorporation of liquid electrolyte solutions into solid polymer matrices to produce "wet" or "plasticized" gel polymer electrolytes (GPEs) is very effective in lowering the ionic conductivity of GPEs to the level of 10^{-3} S/cm at room temperature[16, 17]. The increased ionic conductivity occurs at the expense of mechanical strength of the prepared films. The magnitude of the mechanical strength reduction depends mainly on the type of polymer, the type of solvent and the amount of solvent used. For example, propylene carbonate (PC), an active solvent for linear or branched-chain PEO[16], is a latent solvent for polyacrylonitrile (PAN)[17] and polyvinylidene fluoride (PVdF)[18]. Therefore, if heavily plasticized with PC (~40-60 wt%), dimensionally stable films can be prepared from PAN- and PVdF-based gel electrolytes, but not from linear or branched-chain PEO-based gel electrolytes. This is because active solvents will dissolve polymers at room temperature, whereas latent solvents will form gels with polymers at room temperature, although the gelled polymers may turn into liquid state again at elevated temperatures.

Among the GPEs investigated for use as separators in Li-Ion cells, PAN-, and PVdF-, and polyacrylate-based gel systems have attracted most R&D efforts[16, 18, 19]. Dimensionally stable films have been made from these GPEs containing as much as 80 wt% liquid solvent[17]. Although liquid solvents in these gels are immobilized macroscopically, they still behave much like liquid molecules at the microscopic level. It has been reported that the reactivity of gelled PAN-PC-EC-LiClO$_4$ electrolyte has similar behavior towards lithium metal to that of liquid PC-EC-LiClO$_4$ electrolyte[16].

Hooper [20,21] was able to obtain acceptable conductivity with plasticized PEO films but suffered from poor cycling with lithium metal electrodes. This is typical of the behavior of lithium in liquid organic electrolytes. Shackle et al[22] developed a plasticized radiation cross-linked acrylate electrolyte system with excellent conductivity for lithium systems. The low cycling efficiency of lithium coupled with the tendency of cycled lithium to form lithium dendrites are well known problems for cells with a liquid electrolyte and those with GPE separators. This difficulty is largely responsible for the shift away from lithium metal cells toward the Li-Ion technology with GPE constructions.

Of the three types of polymers mentioned above, PVdF-based polymers have attracted most R&D work of GPE Li-Ion cells because they readily swell in carbonate solvents, possess good chemical and electrochemical stability, and have good mechanical processability. Feuillade and Perche observed in 1975[23] that P(VdF-HFP), the copolymer of vinylidene fluoride (VdF) and hexafluoropropylene (HFP), possessed good ionic conductivity (5×10^{-4} S/cm) and good mechanical strength after swollen in a PC solution of NH_4ClO_4. In the early 1980s, Tsuchida and coworkers also reported that ionic conductivity of PVdF, P(VdF-TFE)[1] and P(VdF-TrFE)[1] polymers, when gelled with some liquid electrolytes (e.g. $LiClO_4$ dissolved in PC), reached 10^{-5} to 10^{-6} S/cm at room temperature[24]. The ionic conductivity of these GPEs was low because only 21.9 to 35.2 wt% solvents were incorporated into the polymer matrices, rather than the 50-60 wt% level needed to reach 10^{-3} S/cm at room temperature. Gozdz and coworkers at Bellcore reported that P(VdF-HFP)[25, 26] and P(VdF-CTFE)[27] polymers[1], after swollen in carbonate-based electrolyte solutions, had very good ionic conductivity ($\sim 10^{-3}$ S/cm) and good mechanical strength at room temperature. These copolymers dissolve in carbonate solvents only under some extreme conditions[28, 29].

Current GPE Li-Ion batteries are flat in format and packaged in thin plastic-foil pouches. A single battery often contains multiple bicells stacked together. Cells using P(VdF-HFP) and P(VdF-CTFE) copolymer as binders, and/or separator materials, have been reported to have good performance characteristics[25] and are currently under active development by some battery manufacturers. We have recently compared the manufacturing processes and performance characteristics of GPE Li-Ion cells with those of liquid electrolyte Li-Ion cells[30].

In this paper, we report our experimental results of the effects of HFP concentration, solvent type, temperature and salt on the swelling behavior of the P(VdF-HFP) in electrolyte solutions, since the cell performance depends, to a large extent, on the amount of electrolyte solution present in GPEs. We then demonstrate that Li-Ion coin cells using P(VdF-HFP) separator can have rate capability equal to or better than those using a microporous separator

---

[1]TFE-tetrafluoroethylene; TrFE-trifluoroethylene; CTFE-chlorotrifluoroethylene.

when appropriate GPEs are selected. Finally, we will discuss several developments in the early stages of commercialization.

PvDF Polymer Experiments

P(VdF-HFP) polymers containing 0%, 8% and 15% of HFP by weight were used for electrolyte solution uptake study. The PVdF polymer samples were from Solvay. The pure homopolymer had the designation 1010, the 8% HFP sample had the designation 20810 and the 15% HFP had the designation 21510. Polymer samples with a diameter of 18 mm were punched out of 2 mm. thick polymer plaques and immersed in 1 M $LiPF_6$ electrolyte solutions at $21^{\circ}C$ or $55^{\circ}C$. The solvents for the electrolyte solutions were DMC, DEC, EC/DMC (1:0.82 by weight), EC/DEC (1:0.82 by weight), and EC/DMC/DEC (1:2:2 by weight), respectively. After immersion in the electrolyte solution for 1, 7, 14 and 30 days, the polymer samples were removed from the electrolyte solutions contained in poly(propylene) bottles, wiped dry and weighed. All experimental procedures, including cell assembly, were carried out inside a dry glove box. The amount of electrolyte solution absorbed by a polymer sample is $(W_b - W_a)$ and the percentage of the weight increase of the polymer sample is calculated from $100*(W_b-W_a)/(W_a)$. Here, W_a and W_b are the weight of the polymer samples before and after immersion, respectively.

Flooded Li-Ion coin cells were built using the 2016 coin cell hardware, 3.2 mil thick $LiCoO_2$ cathode electrodes and 3.6 mil thick MCMB anode electrodes. The binder for both the cathode and the anode electrodes is a PVdF homopolymer. Because the rate capability of GPE Li-Ion cells is much affected by the GPE separators, a design of experiments was carried out for five variables - solvent type, 55 $^{\circ}C$ pre-conditioning, filler type, polymer type, and separator thickness – to explore the relative importance of these variables in affecting the discharge rate capability at room temperature. The preparation of the GPE separator followed the method described in reference[32]. Methanol was used to extract dibutyl phthalate (DBP), a processing and performance-enhancing agent, from P(VdF-HFP). The Li-Ion coin cells were first charged to 4.1 V and then discharged to 3.0 V at C/10 rate at room temperature. The discharge rate capability was determined from subsequent discharge to 3.0 V at C/10, C/5, C/2 and 1C rates and at room temperature. Before each discharge, cells were charged to 4.1 V using constant current (C/2 rate) and then held at 4.1 V until the current dropped to 10% of the C/2 current level.

Results and Discussion

Fig. 1 shows the percentage of weight gain of three P(VdF-HFP) polymer samples immersed in 1 M $LiPF_6$ solutions of DMC, DEC, EC/DMC, and EC/DEC at room temperature. It can be seen from Fig. 1 that increasing the co-monomer content in P(VdF-HFP) increased the degree of swelling of the polymer samples studied. The increased absorbency of the electrolyte solution by the copolymers is related to the lower crystallinity level in the copolymers. It was reported that the crystallinity decreased from 38% for the PVdF

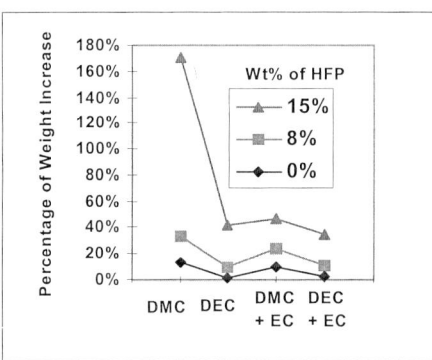

Fig. 1: Percentage of weight increase of P(VdF-HPF) samples soaked in 1 M $LiPF_6$ solutions of DMC, DEC, DMC/EC, and DEC/EC for 30 days at 21 °C.

homopolymer to 27-29% for P(VDF-8%HFP) and to 22% for P(VDF-15%HFP)[28,29]. The effect of solvents on the polymer swelling can also be seen from Fig. 1. When the solvent was changed from DEC to DMC, the electrolyte solution absorbed by the polymer samples increased from 1 wt% to 13 wt% for the PVdF homopolymer and from 41 wt% to 170 wt% for the P(VdF-15%HFP) copolymer. The swelling data of P(VdF-HFP) in pure EC-containing electrolyte is not available at room temperature (21°C) because EC's melting temperature is around 37°C. Therefore, the effect of EC on the swelling of P(VdF-HFP) was estimated from the weight gain of P(VdF-HFP) in electrolytes using EC/DMC and EC/DEC solvent mixtures. It is clear from the data in Fig. 1 that EC is similar to DEC but poorer than DMC as a solvent for swelling P(VdF-HFP) because the trend of solvent uptake by P(VdF-HFP) is DMC>EC/DMC and DEC≈EC/DEC.

Increasing the temperature from 21°C (room temperature) to 55°C substantially increased the swelling of P(VdF-HFP) in the carbonate electrolytes (Fig. 2). The amount of 1M $LiPF_6$-EC/DEC electrolyte absorbed by PVdF homopolymer increased from 2 wt% to 11 wt% when the temperature was raised from 21°C to 55°C. For P(VdF-15%HFP), an increase of the electrolyte absorbency (from 35 wt% to 75 wt%) resulted by this temperature increase. It is interesting to notice that a temperature "memory effect" was observed for these swollen polymers. After the swollen polymers were cooled from 55°C to 21°C and stored for 140 days, less than ~5-10 wt% of the total absorbed electrolyte was excuded from the swollen polymers, to the surrounding electrolytes.

GPE Li-Ion cells may operate at temperatures higher than 55°C in actual use. Therefore, it is necessary to carefully evaluate the mechanical integrity of P(VdF-HFP) copolymers in carbonate electrolytes at elevated temperatures because polymers may loose their mechanical strength or dissolve in carbonate electrolytes if there is sufficient solvent in contact with them. Recently, Meunier[28] reported that almost all P(VdF-HFP) copolymers studied by him, dissolved in EC/DEC, EC/DMC/DEC, EC/PC/DMC, and EC/PC/DEC solvents at 87°C.

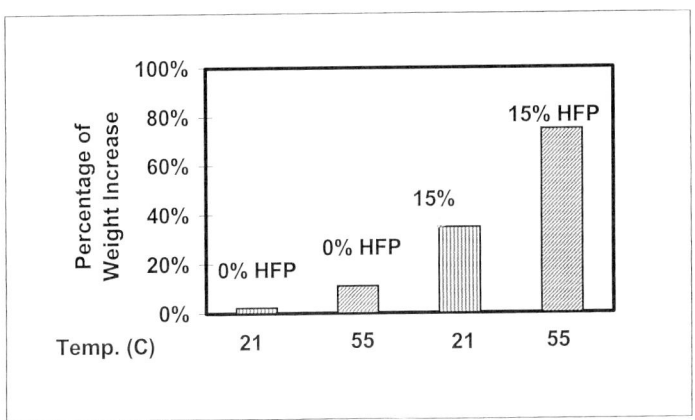

Fig. 2: Effect of temperature on the swelling degree of P(VdF-HFP) in 1 M LiPF$_6$ EC/DEC electrolyte solution.

Fig. 3 compares the swelling behavior of selected P(VdF-HFP) in pure carbonate solvents with that in LiPF$_6$ containing carbonate electrolytes. The data for the pure solvents were taken from reference[26]. The homopolymer exhibited a similar degree of swelling in both the pure solvents and the LiPF$_6$-containing electrolyte solutions, showing no salt effect. A significant salt effect on polymer swelling existed in some cases when the HPF comonomer reached 15% by weight and the solvents used were DMC, EC/DEC and EC/DEC/DMC. The presence of LiPF$_6$ (1M) in the solvents reduced the weight increase of P(VdF-15%HFP) from

86% to 35% if the solvent was EC/DEC or from 276% to 99% if the solvent was EC/DMC/DEC. At 55°C, P(VdF-15%HFP) dissolved completely in pure EC/DEC and EC/DEC/DMC solvents but still possessed the initial circular shape in the corresponding 1 M LiPF$_6$ electrolyte solutions.

Although Li-Ion cells using SPE separators have low discharge rate capability at room temperature, cells using highly conductive GPE separators have at least an equivalent discharge rate capability as that of Li-Ion cells using microporous polyethylene separators. This is because the ionic conductivity of a GPE film can be higher than that of a microporous polyolefin separator immersed in a liquid electrolyte. Fujita and coworkers reported that the room temperature ionic conductivity for a microporous polyolefin separator immersed in 1M LiPF$_6$-EC/DMC/DEC is about 8×10^{-4} S/cm which is lower than $1-3 \times 10^{-3}$ S/cm reported for some GPEs[32]. It may be that the internal surface charge can enhance the conductivity.

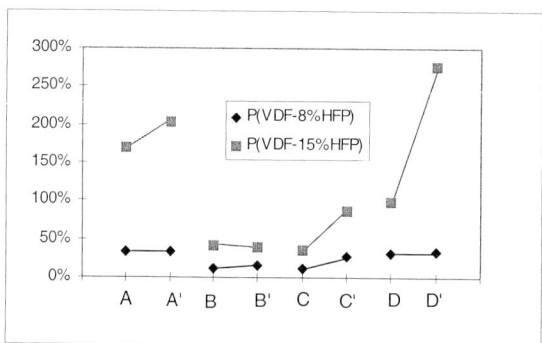

Fig. 3: Percentage of weight increases of P(VdF-HFP) in 1 M LiPF$_6$ electrolyte solutions and pure solvent solutions at room temperature. A' stands for DMC, B' for DEC, C' for EC/DMC, and D' for EC/DMC/DEC. A, B, C and D are 1 M LiPF$_6$ in A', B', C' and D', respectively.

The good discharge rate capability of a Li-Ion cell using a polymer electrolyte film can be seen in Fig. 4 which displays the voltage profiles of a coin cell discharged at rates C/5, C/2 and 1C, at room temperature. The separator employed in the cell was a P(VdF-HFP) composite film. The 1C discharge capacity is 92% of that discharged at C/5. Replacing the PVdF separator with a conventional microporous polyethylene separator reduced the 1C discharge capacity to 62% of the C/5 capacity. This difference may be related to a contact resistance at the electrode-polyethylene separator interface in the coin cell construction.

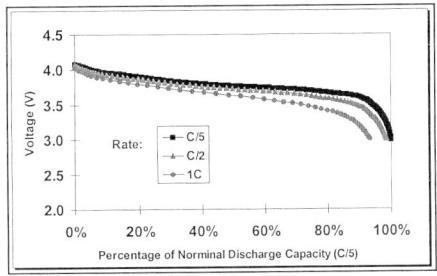
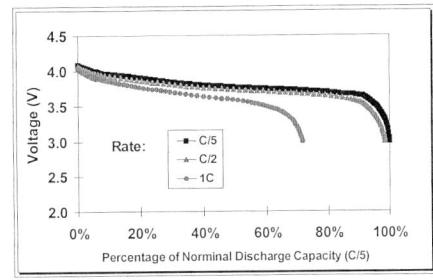

Figure 4. Discharge profiles of a flooded
Li-ion coin cell using a GPE separator.

Figure 5. Discharge profiles of a flooded
Li-ion coin cell using a microporous
polyethylene separator.

Because the cell discharge rate capability of a GPE Li-Ion cell using thin electrodes is closely
related to the conductance of the GPE separator use, a variable search experiment was
designed to find factors (or parameter changes) that have a large influence on cell discharge
rate capability at room temperature. Five factors–solvent type (A), high temperature
formation treatment (B), silica vs. alumina separator filler type (C), HFP content polymer type
(D), and separator thickness (E) were considered in this experiment. A 1 M $LiPF_6$-
EC/DMC/DEC electrolyte, 55°C pre-conditioning, SiO_2 filler, P(VdF-15%HFP), and 2 mil
thick separator were used in the construction of cells that gave the best performance, whereas,
cells made with 1M $LiPF_6$ EC/DEC electrolyte, no high temperature treatment, Al_2O_3 filler,
and P(VdF-11%HFP) gave the worst performance.

Among all parameter changes considered in this experimental design, changing temperature
(variable B) from 21 °C to 55°C is the most effective way for increasing high rate discharge
capability. Changes in variables C, D and E in the specified ranges have moderate effects on
discharge rate capability. Replacing an EC/DEC-based electrolyte by an EC/DEC/DMC-
based electrolyte seemed to be less effective than other changes on affecting the rate
capability.

Commercialization Efforts and the "Bellcore" Process

The greatest activity has concentrated on developing and commercializing the Bellcore (now
Telcordia) process based on a PVdF plastic electrolytes. The embodiment of the PVdF based
battery system has been described by Tarascon et al[25, 26, 27, 31]. This has become known as the
Bellcore process. A schematic of the Bellcore process is given in Fig. 6. Films of the anode,

238

cathode and separator are cast using PVdF polymer as binder and electrolyte materials. The essential element of the process is the incorporation of dibutyl phthalate (DBP) as a plasticizer in the manufacture of the electrode films. The DBP plasticized films allow easy

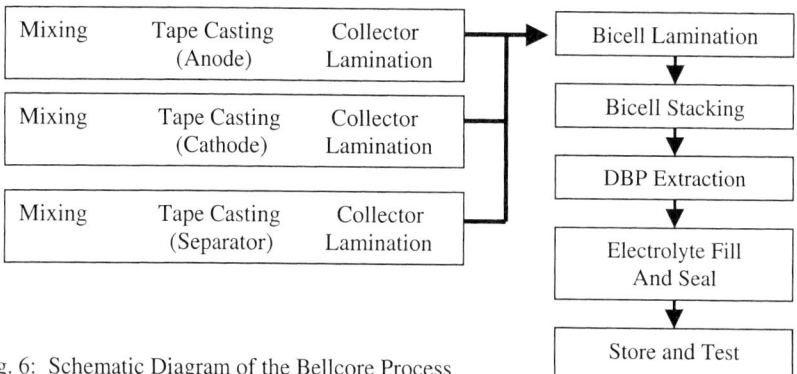

Fig. 6: Schematic Diagram of the Bellcore Process

handling and processing of the laminate materials into finished cells. The DBP also functions as a pore-former for the absorption of electrolyte into the polymer. The DBP must be completely extracted from the polymer material, before the electrolyte is added to the assembled cell. Failure to reduce the concentration of DBP below 350 ppm results in poor cycle life for the finished product[33]. DBP is electrochemically active and must be completely removed before the cell is activated by addition of the electrolyte. The removal of DBP creates minute fissures and voids in the solid PVdF that are filled by the addition of electrolyte during cell activation.

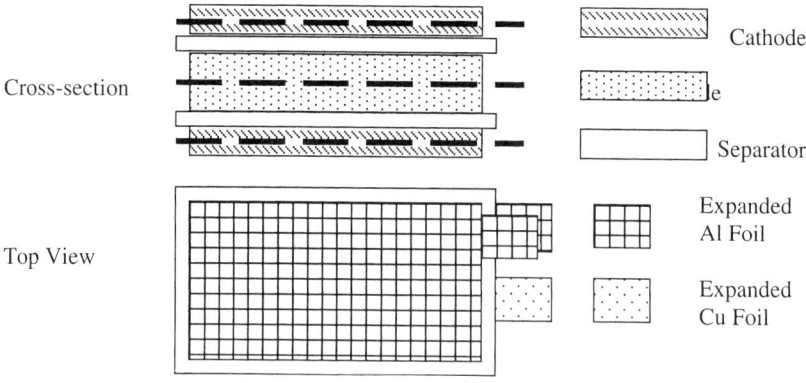

Fig. 7: Typical construction of a polymer Bicell. The separate layers and current collectors are heat laminated together

Typical construction of the polymer cell is shown in Fig. 7. This construction with two electrodes on each side of a center electrode, is termed a "bicell." These are the basic units of construction of complete batteries and may be combined to produce higher capacity cells. Polymer batteries can be fabricated in sheets that result in large footprint batteries. Since the same polymer is used throughout the active components, the polymer effectively bonds the cell components together as a single unit. There are no phase boundaries between the active cell components and the separator to introduce unwanted resistance. Consequently, the cells do not require application of external pressure for good operation. The PVdF electrolytes are very stable and can accommodate the 4-volt nickel, cobalt and manganese cathode systems. The PVdF electrolytes are reported to be "transparent" to the chemistry.

The performance characteristics of the Bellcore type Li-Ion cells are illustrated in Figs:8, 9 and 10. The cells are constructed as in Fig. 7. The cathode is a manganese spinel, $Li_{1+0.05}Mn_2O_4$, the anode is graphite and the electrolyte is 1 M $LiPF_6$ in a 50:50 mixture of ethyl carbonate and dimethyl carbonate solvents. Excess lithium is incorporated into the manganese spinel cathode material to stabilize its crystal structure and enhance long term operation. The cells have good rate capability and cycle life. The discharge curve in Fig. 8,

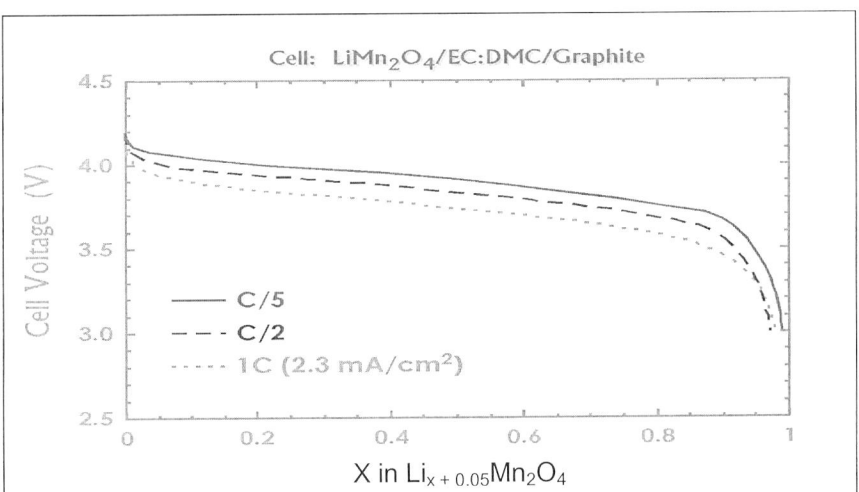

Fig. 8: Discharge Curve of a Bellcore technology cell at the five hour (C/5), 2 hour (C/2) and one hour (1C) rates. (Courtesy of Bellcore (now Telcordia))

shows little difference between the capacity delivered at a five hour (termed C/5) discharge rate and the capacity of the cell when discharged at the one hour (termed 1C) rate discharge.

This performance is very similar to that for a liquid electrolyte Li-Ion cell. The discharge curve is fairly flat with an average discharge voltage of 3.7 volts. The manganese cathode has about a 0.1-volt higher discharge voltage than does the lithium cobalt oxide cathode of the liquid electrolyte cells.

The cycle life for the Bellcore cells is illustrated in Fig. 9. The cells are discharged at a three hour (C/3) rate and then recharged using a current limited constant potential regime. Here the current is limited to the two hour rate. As the cell approaches full charge, the voltage limit forces the current lower. The charge is stopped when the charge current falls to 10% of its

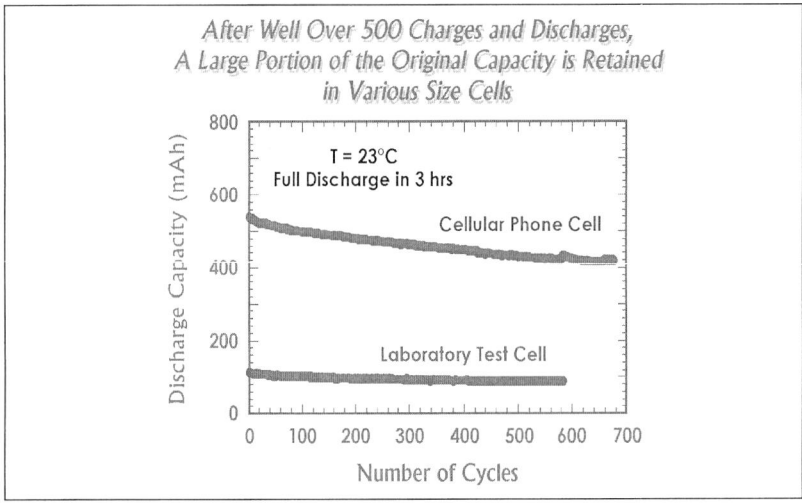

Fig. 9 Cycle Life of 2 in. X 3 in. X 0.032 in. bicells at the two hour discharge rate. (Courtesy of Bellcore)

initial value. The laboratory and cellular phone cells show acceptable cycle life. The capacity falls slowly and reaches 80% of their initial capacity after more than 500 cycles.

Thomas and Betts (High Energy Technology) and Ultralife have introduced polymer batteries based on Bellcore-type technology. Several others including Valence Technology, have also been developing polymer batteries based on this technology. All told, there are over 20 licensees of the Bellcore technology worldwide.

Polymer cells can handle the high current pulses of the GSM cellular phone application. Fig. 10 shows the performance comparison of the polymer cell with that for a Ni-Cd cell of similar capacity on a simulated GSM protocol with 95 millisecond pulses for both transmission and receive. The 3C rate simulates the 1.5 ampere GSM transmission pulse. Because of the difference in unit cell voltages, two Li-Ion cells are compared with 5 Ni-Cd cells connected in series. The two systems gave essentially the same performance. There is little difference between the high current transmit and standby performance of each system. This difference in voltage between transmit and receive is directly related to the internal

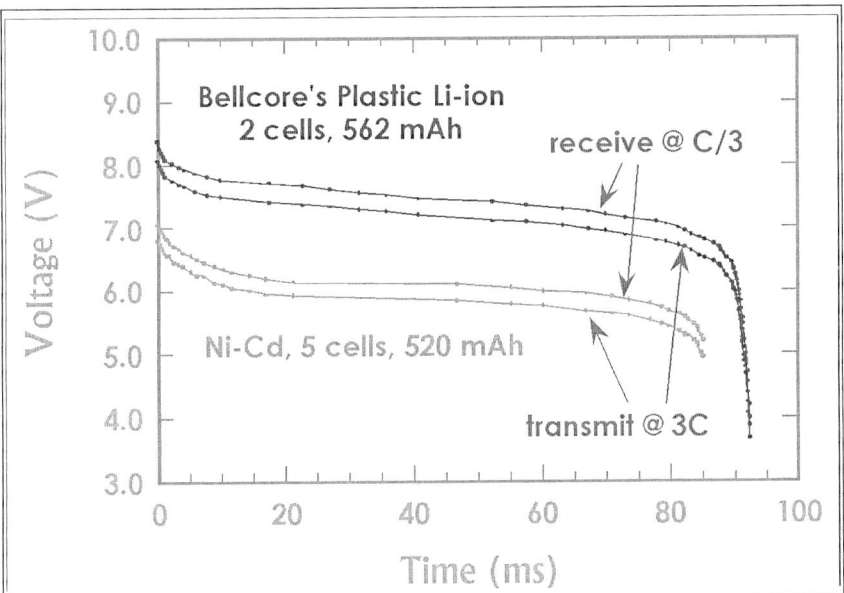

Fig. 10: Comparison of the characteristics of a 562 mAh, two-cell polymer Li-Ion and 520 mAh five cell nickel cadmium battery on a simulated GSM discharge regime. (Courtesy of Bellcore (now Telcordia))

resistance of the battery. Both batteries have similar internal resistance. Overall, the polymer cells have demonstrated equivalent performance to that of a typical Li-Ion liquid electrolyte cell.

The field of polymer batteries is a very active technical area. There are many efforts to develop Li-Ion polymer cells in various stages of development. It is estimated that over 30 different companies have development efforts that can lead to commercial production. Battery Engineering has introduced a Li-Ion battery based on acrylate electrolyte[34, 35]. The electrolyte is thermally polymerized in situ after the cell is assembled of three monomer moieties. A plastic mesh is embedded in the electrolyte layer to impart mechanical strength to the polymer. The cells do not contain a volatile solvent as the boiling point of the plasticizer is 210°C. The flexible polymer chain provides free volume for good ion transport and conductivity.

Saft America has announced a Li-Ion polymer cell using a similar approach but has not disclosed the identity of the polymer electrolyte[36, 37]. The polymer is based on PVdF and consists of two different interpenetrated phases: the swollen polymer and free liquid. The polymer electrolyte is prepared by a phase inversion process and has good conductivity. The process results in a separator material with a 60% void volume and pore size of 0.1m to 1.0 m. The polymer swells about 20% to 30% by weight based on electrolyte absorption.

Electrofuel in Canada announced a 160 Wh thin flat battery for notebook computers that takes advantage of the thin flat format possible with polymer cells[38]. The battery measures 220 mm. X 206 mm. X 9.5 mm. and weighs 0.999 kg with a titanium case. This translates to 250 Wh/l and 160 Wh/kg. The fully charged battery is said to be able to power a notebook computer for about 8 hours. The identity of the polymer electrolyte was not revealed. The battery was enclosed in a titanium case and attaches directly to the bottom of a notebook computer. The company claimed that energy densities of 175 Wh/kg and 475 Wh/l were possible.

Several Japanese companies including Sony, Sanyo, Matsushita and A&T are now actively marketing polymer Li-Ion batteries. At this point in time, very little is known about the technology basis for these developments.

Conclusion

Given proper internal cell design, polymer Li-Ion cells exhibit equivalent performance to that of the regular liquid filled Li-Ion cells. They may have a weight advantage given the use of polymer foil packaging materials. Polymer Li-Ion cells are internally bonded and do not need external pressure to maintain good contact between the electrodes and the separator. Polymer Li-Ion batteries are available from several suppliers based on a variety of technologies and polymer systems. The absorption of 1 M $LiPF_6$ electrolyte solutions by P(VdF-HFP) polymers depends on the concentration of HFP unit in the polymer, solvent type, and temperature. Increasing the HFP unit content in P(VdF-HFP) and temperature increased the swelling degree of the polymers considerably. The amount of the electrolyte solution absorbed by P(VdF-HFP) at room temperature showed the following trend: DMC>EC/DMC>DEC≈EC/DEC. The presence of $LiPF_6$ in carbonate solvents did not affect the degree of swelling of PVdF and P(VdF-8%HFP) in all solutions studied but lowered the swelling degree of P(VdF-15%HFP) in DMC, EC/DEC and EC/DMC/DEC solution at room temperature or prevented dissolution in the them at 55 °C. The swelling and memory effect related to gel formation at higher temperatures may limit the use of batteries using PVdF materials to 55°C or less.

Acknowledgements

D. J. Allessio and S. E. Osmialowski carried out the experiments and summarized some of the data used in this paper. The authors also gratefully acknowledge Dr. Vassillis Keramidas of Bellcore for supplying the performance curves for their polymer cells.

References

1. A. A. Blumberg, S. S. Pollack and C. A. J. Hoeve, *J. Polym. Sci. Part A* **2**, 2499 (1964)

2. P. V. Wright, *J. Br. Polym.* **7**, 319 (1975)

3. D. E. Fenton, J. M. Parker and P. V. Wright, *Polymer* **14**, 589 (1973)

4. M. B. Armand, J. M. Chabogno and M. Duclot, in *Fast Ion Transport in Solids*, P. Vashista, J. N. Mundy and G. K. Shenoy (Eds.), Elsevier Applied Science, New York, 1979, p. 131ff

5. P.G. Bruce and C.A Vincent, *J. Chem. Soc. Faraday Trans.* **89**, 3187 (1993)

6. F. M. Gray, *Solid Polymer Electrolytes, Fundamentals and Technological Applications*, VCH Publishers, New York, 1991

7. J.S. Tonge and D.F Shriver, in *Polymers for Electronic Applications*; J.H. Lai (Ed.), (CRC Press Inc., Florida, 1989), p. 157.

8. K.M. Abraham, *Electrochem Acta* **38**, 1233 (1993).

9. C.V. Nicholas, D.J. Wilson, C. Booth and J.R.M. Files, *Brit. Polym. J.* **20**, 289 (1988).

10. P.M. Blonsky, D.F. Shriver, P. Austin and H.R. Allcock, *J. Am. Chem. Soc.* **196**, 6854 (1984)

11. K. Nagaoka, H. Naruse, I. Shinohara and M. Watanabe, *J. Polym. Sci. Polym. Lett. Ed.* **22**, 659 (1984).

244

12. J.M.G. Cower and A.C.S. Martin, *Polymer* **28**, 627 (1987).

13. M. A. Armand, W. Gorecki and R. Andreani, in *Proc. Second Intl. Mtg. on Polymer Electrolytes,* B. Scrosati (Ed.), Elsevier, New York, 1990, p. 91.

14. C.A. Angell, C. Liu and E. Sanchez, *Nature* **362**, 137 (1993).

15. F. Capuano, F. Croce and B. Scrosati, *J. Electrochem. Soc.* **138**, 1918 (1991).

16. M. Alamgir and K.M. Abraham, in *Lithium Batteries: New Materials Developments and Perspectives,* G. Pistoia (Ed.), Elsevier Science B.V., Amsterdam, The Netherlands, (1994), p.93.

17. K.M. Abraham and M Alamgir, *J. Electrochem. Soc.* **137**, 1657 (1990).

18. E. Tsuchida, H. Ohno and K. Tsunemi, *Electrochim. Acta,* **28**, 591 (1983).

19. R. Koksbang, I.I. Olsen and D. Shackle, *Solid State Ionics* **69**, 320 (1994).

20. A. Hooper, 2nd International Meeting on Lithium Batteries, Paris, April 1984, Poster Presentation.

21. A. Hooper and J. M. North, U.S. Patent 4,547,440, Oct. 15, 1985

22. M-T Lee, D. Shackle and G. Schwab, U.S. Patent 4,8330,939, May 16, 1989.

23. G. Feuillade, Ph. Perche, *J. Appl. Electrochem.* **5**, 63 (1975).

24. E. Tsuchida, H. Ohno and K. Tsunemi, *Electrochim. Acta*, **28**, 591 (1983).

25. A. S. Gozdz, J. -M. Tarascon, O. S. Gebizlioglu, C. N. Schmutz, P. C. Warren and F. K. Shokoohi, in *Rechargeable Lithium and Lithium-Ion Batteries*, S. Megahed, B.M. Barnett and L. Xie (Eds.), *(The Electrochem. Proceedings.*, **94-28**, 1994), p. 400ff.

26. A.S. Gozdz, C.N. Schmutz, J.-M. Tarascon and P.C. Warren, U.S. Patent 5,456,00 (Oct. 10, 1995).

27. A.S. Gozdz, C.N. Schmutz and P.C. Warren, U.S. Patent No. 5,571,634 (Nov. 5, 1996).

28. V. Meunier, paper presented at *The 15th International Seminar & Exhibit on Primary & Secondary Batteries*, Fort Lauderdale, March 2-5, 1998.

29. M. Burchill, in *Advances in Batteries and Solar Cells*, JEC Press, Nagoya, 1998, Vol. 17., page 144.

30. W. Huang and J.R. Akridge, in *Proc. of 33rd Intersociety Energy Conversion Engineering Conference*, Colorado Springs, CO, August 2-6,1998, paper I102.

31. J.-M. Tarascon, A.S. Gozdz, C. Schmutz, F. Shokoohi and P.C. Warren, *Solid State Ionics* **86-88**, 49 (1996).

32. Y. Fujita, in *Symposium on Possible Applications of Polymer: High-Tech Lithium Battery Supported with Polymer,* Polymer Society of Japan, Tokyo, September 18-19, 1997, p. 60.

33. B. Lee, C. Yi, J. S. Kim and J. T. Kim, Paper presented at *Korean Electrochemical Society*, October, 1999

34. L. Sun, Y. Xu, M. Albu, S. Iqbal and Y. Uetani, in *Proceedings of the 38th Power Sources Symposium*, IEEE, New York, 1998, p274

35. L. Sun and I. Irving, in *Proceedings of Power 98*, Giga Information Group and Arthur D. Little, San Jose, October 4 – 6, 1998.

36. X. Andrieu, C. Jehoulet, F. Boudin and I. Olsen, in *Proceedings of the 38th Power Sources Symposium*, IEEE, New York, 1998, p266

37. I. Olsen, in *Proceedings of Power99*, Giga Information Group and Arthur D. Little, San Jose, October 3 – 6, 1999.

38. S. Das Gupta, J. Jacob and M. Murdoch, in *Proceedings of the 16th International Seminar and Exhibit on Primary and Secondary Batteries*, Ft. Lauderdale, FL, March 1 – 4, 1999.

*Macromol. Symp. **159**, 247–257 (2000)*

Resorbable Initiators for Polymerizations of Lactones

Hans R. Kricheldorf, Ingrid Kreiser-Saunders, Dirk-Olaf Damrau

Institut für Technische und Makromolekulare Chemie, Universität
Hamburg, Bundesstr. 45, D-20146 Hamburg, Germany

SUMMARY: Numerous nontoxic (resorbable) salts were prepared from
cations and anions belonging to the human metabolism, such as Na, K,
Mg, Ca, Zn and Fe in combination with chloride, iodide, hydroxide,
carbonate, acetate, stearate, glycolate, L-lactate, D-mandelate and various
N-substituted α-amino acids. All these salts were used as initiators for
polymerizations of L-lactide in bulk at 100 - 180°C. Furthermore,
Grignard reagents, hemin and hematin were included in this study. Zn L-
lactate was found to be the most useful initiator in terms of reactivity,
maximum molecular weight of the isolated poly(L-lactide) and its optical
purity. Zn L-lactate initiated copolymerizations of L-lactide and glycolid
or L-lactide and ε-caprolactone were also studied. Finally, first studies of
the polymerization mechanism of zinc stearate and zin 2-ethylhexanoate
were performed. They suggest that zinc carboxylates combined with an
alcohol as coinitiator form reactive zinc alkoxides which are the true
initiators at temperatures < 150°C.

            CCC 1022-1360/00/$ 17.50+.50/0

Introduction

Biodegradable polymers (mainly polyesters) have recently attracted increasing interest for a variety of applications. The broadest interest and usefulnees is attributed to polylactides and copolyesters of lactic acid. These polyesters can be produced either by ring-opening polymerization of lactides (four stereoisomers are known) or by direct polycondensation of lactic acid.

Regardless of the procedure good transesterification catalysts are needed to obtain sufficiently high molecular weights. The most widely used transesterification catalysts are tin salts particularly Sn(II)2-ethylhexanoate $(SnOct_2)$[1,2]. This catalyst is attractive because it is highly reactive, and because it yields high molecular weight polylactides without racemization. Furthermore, it has been accepted as food additive by the American FDA. The reason for this application is its cytotoxicity (which is typical for most tin compounds) against almost any kind of microorganism, so that it plays the role of a food stabilizer. The now envisaged production of several hundred thousands tons of polylactide per year raises the concern that the delivery of larger quantities of poisonous tin compounds to the environment and to the human body (in the case of medical or pharmaceutcal applications) is not tolerable.

Another frequently used group of initiators are aluminum alkoxides or complexes. Aluminum ions are certainly less toxic than tin ions, but they do not belong to the human metabolism and are suspicious to support the Alzheimer desease. This situation prompted us 15 years ago[3-12] to start a systematic search for a non-toxic, resorbable initiator, and the present paper provides a short review of our results.

Results and Discussion

The strategy used for finding a resorbable initiator is based on the simple consideration that they should consist of components (e.g. cations and amions) familiar to the human metabolsm. The cations meeting this requirements are:

$$Na^{\oplus}, K^{\oplus}, Mg^{2\oplus}, Ca^{2\oplus}, Zn^{2\oplus}, Fe^{2\oplus}$$

The human body also contains traces of $Mn^{2\oplus}$, but manganese salts show a significant toxicity and were found to be poor polymerization catalysts[5]. Therefore, this review does not pay further attention to $Mn^{2\oplus}$ salts.

The number of species in the human body which can play the role of counterons (anions) is far higher than the number of cations. The most important examples are:

chloride and iodide
oxide, hydroxide and carbonate
acetate and higher fatty acids
lactate, tartrate, citrate
α-amino acids and peptides

Numerous salts and complexes were prepared (or purchased) from the aforementioned cations and anions, and evaluated with regard to their usefulness as initiators. Most experiments were conducted in such a way that the dry and finely powdered initiators were mixed with recrystallized L-lactide and heated in bulk to temperatures in the range of 100 - 180°C. The usefulness and efficiency of the potential initiators were defined by the reactivity, the maximum molecular weights and the optical purity of the isolated poly(L-lactide)s.

The results of a large number of experiments were compiled in Tables 1 - 3.

Table 1.   Efficiency of metal halides as polymerization catalyst for L-lactide

Cation	Cl	Br	I	Comment
Na	---	---	---	inactive
K	---	---	---	inactive
Mg	---	---	---	inactive
Ca	---	---	---	inactive
Fe	+	+	+	active
Zn	+	+	+	active

Table 2.   Efficiency of basic metal salts as polymerization initiators for
L-lactide

Cation	O^{2-}	OH-	CO_3^{2-}	Comment
Na	---	---	---	
K	---	---	---	Strong
Mg	---	---	---	Racemiz.
Ca	---	---	---	and
Fe	---	---	---	Chain
Zn	---	---	---	Transfer

Table 3.   Efficiency of zinc salts of falty acid and hydroxy acids as polymerization initiators for L-lactide

Cation	Acetate	Stearate	Glycolate	Lactate	Comment
Na	---	---	---	---	Deprot.
K	---	---	---	---	Racem.
Mg	---	---	---	---	Chain
Ca	---	---	---	---	Transfer
Fe	---	---	---	---	
Zn	+	+	+	+	useful

In connection with the evaluation of magnesium and calcium halides[4,5] (Tab. 1) it should be mentioned that also alkyl Grignard reagents were tested[8]. Due to the elimination of alkanes (resulting from the deprotonation of lactide) even Grignard reagents may be classified as resorbable initiators. However, they only showed a moderate performance in terms of reactivity and molecular weight. When used in bulk polymerizations, they caused racemization. $Zn^{2\oplus}$, $Fe^{2\oplus}$ and $Mn^{2\oplus}$ halides proved to be reactive enough to initiate polymerizations of L-lactide at temperatures $\geq 150°C$[9,10], but only $ZnBr_2$ gave satisfactory results. When compared to Zn lactide ($ZnLac_2$), $ZnBr_2$ has the short-comings of a greater hygroscopicity, and the bromide ion is not part of the human metabolism, even though the toxicity of small amounts is extremely low.

When the oxides, hydroxides and carbonates were studied (Tab. 2) it was found that all these basic salts are useless regardless of the cation. The reasons for this classification are low to moderate molecular weights and signifcant racemization. Both effects are interconnected. Lactides are relatively acidic monomers having pK_a values of the α-protons around 17 ± 2. The deprotonation (eq. 1) generates a planar anion due to delocalization of the negative charge. The reprotonation of this anion can occur from both sides of the ring resulting in racemization. Furthermore, the lactide anion can initiate a new chain growth, so that deprotonation produces high conversion but low molecular weights. Therefore, it is an important result of this study that those initiators favoring racemization also tend to yield lower molecular weights. The influence of $FeLac_2$ on the optical purity of unreacted L-lactide recovered from incomplete polymerizations is documented in Table 4. The influence of various organic and inorganic bases on the racemization of L-lactide was also studied in ref.[13].

Table 4.   Influence of FeLac$_2$ on the optical purity of L-lactide upon polymerization in bulk at 150°C (M/I = 1000/1)[a)]

Characterized product	Reaction time h		
	8	24	48
Polylactide isolated after precipitation in Et$_2$O at 20°C	Yield 10 %	48 %	75 %
Oligolactides isolated from the filtrate after extraction with Et$_2$O	Yield 65 %	26 %	15 %
Monomer obtained by evaporation of the Et$_2$O extract	Yield 21 % $[\alpha]_D^{20}$ - 250	18 % - 226	18 % - 188
Starting material: $[\alpha]_D^{20}$	---	---	---

a) The optical rotations were measured in CHCl$_3$ at 20°C with c = 1 g/dL

Table 5.   ZnLac$_2$ initiated polymerization of L-lactide in bulk at 150°C

Mon Init.	Time (h)	Yield (%)	η_{inh}[a)] dL/g	$[\alpha]_D^{20\ b)}$
500	96	91	0.46	-
500	192	90	0.48	- 155
1000	96	90	0.51	-
1000	192	88	0.51	- 157
2000	96	92	0.82	-
2000	192	90	0.66	- 160
4000	96	88	0.89	-
4000	192	91	0.70	- 159
8000	96	81	0.90	-
8000	192	85	0.81	- 158

a)   measured at 20°C with c = 2 g/L in CH$_2$Cl$_2$
b)   measured at 20°C with c = 1 g/L in CHCl$_3$

Table 6. Zn aceturate and Zn-L-prolinate-catalyzed polymerizations[a] of
L-lactide with variation of the monomer/catalyst ratio

Polymer No.	Catalyst	Monomer Catalyst	Yield %	η_{inh}[b] dL/g	$[\alpha]_D^{20}$ [c]
1	Zn-aceturate	1000/1	87	0.44	- 156
2		2000/1	81	0.48	-
3		4000/1	87	0.52	-
4		8000/1	84	0.65	- 157
5	Zn-prolinate	1000/1	88	0.38	- 144
6		2000/1	84	0.44	-
7		4000/1	79	0.48	-
8		8000/1	75	0.62	- 147

a) all polymerizations were conducted in bulk at 150°C (192 h)

b) measured at 20°C with c = 2 g/L in CH_2Cl_2

c) measured at 20°C with c = 1 g/L in $CHCL_3$

$$\text{(1)}$$

When the acetates and L-lactates of Zn, Fe and Mn were compared, the L-lactates of all three cations gave better results in terms of higher molecular weights and less racemization. Furthermore, the Zn lactate was superior to the Zn glycolate or Zn mandelate[9]. Moreover, the L-lactate of Zn proved to be better than the L-lactates of Fe, Mn, M_g and Ca[14]. Finally, it was found that Zn L-lactate yield slightly higher molecular weights and optical rotation than the Zn salts of N-acetylglycine, N-rosyl-glycine, L-proline, L-pyroglutanic acid or N-acetyl-4-aminobenzoic acid[11]. In other words Zn L-lactate proved to be the most attractive and useful resorbable initiator of all our studies (including hemin and hematin which gave poor results[6] None the less, Zn L-lactate is a relatively sluggish initiator and temperatures above 140°C are required for bulk polymerizations of L-lactide. Some typical results obtained with Zn-L-lactate and Zn amino acid salts were summarized in Tables 5 and 6.

A particularly useful property of ZnLac$_2$ as initiator is the combination with an alcohol playing the role of a coinitiator. The coinitiator accelerates the polymerization process slightly and allows a control of the molecular weight via the monomer/coinitiator ratio. In this way each polylactide chains obtain a stable ester endgroup at one chain end, whereas a hydroxy group forms the second chain end after hydrolysis or methanolysis of the Zn-O bond. Such studies were first performed with benzylalcohol as coinitiator[9]. However, an interesting aspect of this strategy is the use of bioactive alcohols or phenols as coinitiator such as geraniol, testosteron, stigmasterol α-tocopherol etc. With low monomer/coinitiator ratios vitamines, hormones or drugs modified with short oligolacitde substitutents were obtained[9] (eq. 2).

In addition to the homopolymerization of L-lactide the homopolymerization of 1,4-dioxane-2-one was studied and high molecular weights were obtained when the polymerizations were performed in bulk at 100°C (eq. 3). However, due to unfavorable thermodynamical properties of this monomers the yields never exceed 70 %[12]. Furthermore, 1:1 copolymerization of L-lactide and glycolide or L-lactide and ε-caprolactone were conducted. When compared to $ZnCl_2$, Zn I_2 or Zn (stearate)$_2$, Zn L-lactate proved to be the most active initiator and transesterifcation catalyst. Although perfectly random sequences were never obtained, the blockiness of the sequences was so low that in all cases copolylactones soluble in CH_2Cl_2 were obtained. Such a good solubility is needed for an application in drug-delivery systems. In summary, Zn L-lactate showed a number of positive properties such as low costs, easy synthesis, good stability on storage, high efficiency as inititor, but it seems to be difficult to achieve number average molecular weights above 10^5.

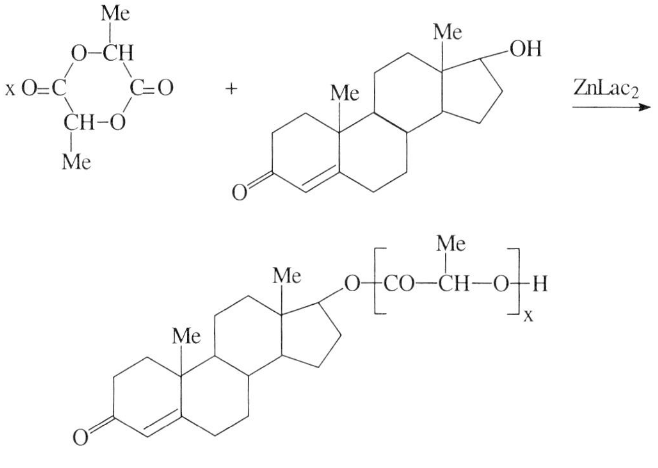

(2)

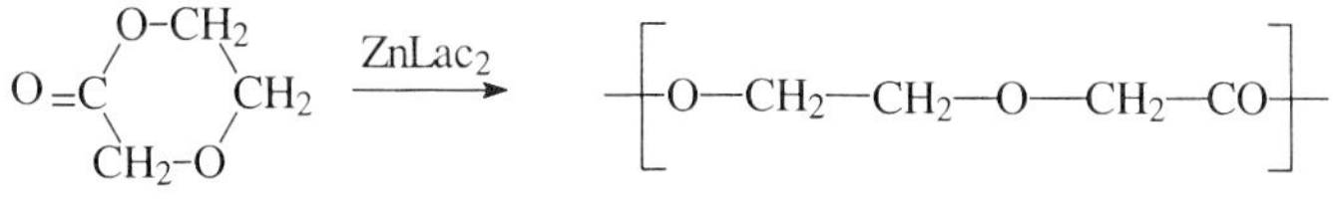

(3)

Quite recently mechanistic studies concerning zinc carboxylates as initiators were started. Zinc stearate and zinc 2-ethylhexanoate (ZnOct$_2$) were used as initiators and (for analytical reasons) benzylalcohol served as coinitiator. First model reactions indicated that the zinc carboxylates react with the alcohol at temperatures $\geq 120°C$ with formation of benzylesters (eq. 4). This esterification is a slow process below 120°C but rapid and nearly quantitative at 180°C a temperature needed for the technical production of poly(L-lactide). The liberation of water generates in the first step Zn-OH groups, but with increasing conversion ZnO precipitates from the reaction mixture of the neat reactants (eqs. (5) + (6)). These results suggest that the intermediately formed zinc hydroxide groups or the zinc-alkoxide groups formed by exchange with an alcohol (eq. (7)) are the active species which react with lactide according to the established coordination-insertion mechanism[15] (eqs. (8)).

$$Zn(O_2C-C_{17}-H_{35})_2 + \longrightarrow \begin{array}{l} Zn(O_2C-C_{17}H_{35})\,OH \\[2mm] C_6H_5\text{-}CH_2O_2C\text{-}C_{17}H_5 \end{array} \qquad (4)$$

$$\begin{array}{l} Zn(O_2C\text{-}CH_{17}H_{35})\,OH \\[2mm] +\ C_6H_5CH_2OH \end{array} \longrightarrow \begin{array}{l} C_6H_5\text{-}CH_2\text{-}O_2C\text{-}C_{17}H_{35} \\[2mm] +\ Zn(OH)_2 \end{array} \qquad (5)$$

$$\downarrow$$

$$ZnO + H_2O \qquad (6)$$

$$\begin{array}{l} Zn\,(OH)_2 + \\[2mm] C_6H_5-CH_2OH \end{array} \xrightarrow{\ -H_2O\ } Zn(OH)\text{-}O\text{-}CH_2\text{-}C_6H_5 \qquad (7)$$

$$XZn-OR \longrightarrow XZn\text{-}O-(A)-CO-O \qquad (8)$$

(A)

References

1. Schindler, R. Jeffcoat, G.L. Kimmel, C.G. Pitt, M.E. Wall, R. Zweidinger in "Contemporary Topics in Polymer Science (E.M. Pearce, J.R. Schaefgen Eds.) Plenum Press, N.Y. 1977 Vol. 2 p. 251
2. H.R. Kricheldorf, I. Kreiser-Saunders, C. Boettcher, *Polymer* **36** 1253 (1995) and literature cited therein
3. H.R. Kricheldorf, J.M. Jonté, M. Berl, *Makromol. Chem. Suppl.*
4. R. Dunsing, H.R. Kricheldorf, Polym. *Bull.* **14** 491 (1985)
5. H.R. Kricheldorf, A. Serra, Polym. *Bull.* **14** 497 (1985)
6. H.R. Kricheldorf, C. Boettcher, *Makromol. Chem.* **194** 463 (1993)
7. H.R. Kricheldorf, S.-R. Lee, *Polymer* **36** 2995 (1995)
8. H.R. Kricheldorf, M. Lossin, *H.M.S.-Pure Appl. Chem.* **A34** 179 (1997)
9. H.R. Kricheldorf, D.-O. Damrau, *Macromol. Chem. Phys.* **198** 1753 1997)
10. H.R. Kricheldorf, D.-O. Damrau, *Macromol. Chem. Phys.* **198** 1767 (1997)
11. I. Kreiser-Saunders, H.R. Kricheldorf, *Macromol. Chem. Phys.* **199** 1081 (1998)
12. H.R. Kricheldorf, D.-O. Damrau, *Macromol. Chem. Phys.* **199** 1089 (1998)
13. H.R. Kricheldorf, I. Kreiser-Saunders, *Makromol. Chem.* **191** 1057 (1990)
14. H.R. Kricheldorf, D.-O. Damrau, *J.M.S. - Pure Appl. Chem.*
15. H.R. Kricheldorf, M. Berl, N. Scharnagl, *Macromolecules* **21** 286 (1988)

*Macromol. Symp. **159**, 259–266 (2000)*

Production of Polyhydroxyalkanoates by Fermentation of Bacteria

Sang Yup Lee*, Jong-il Choi, Seung Hwan Lee

Department of Chemical Engineering and BioProcess Engineering Research Center, Korea Advanced Institute of Science and Technology, 373-1 Kusong-dong, Yusong-gu, Taejon 305-701, Korea. E-mail: leesy@sorak.kaist.ac.kr

SUMMARY: Polyhydroxyalkanoates (PHAs) are carbon and energy reserve material accumulated by numerous microorganisms and have been drawing much attention as biodegradable substitutes for conventional nondegradable plastics and elastomers. There are a number of different PHAs having a variety of material properties based on the different monomer composition. Poly(3-hydroxybutyrate) and poly(3-hydroxybutyrate-co-3-hydroxyvalerate) are now efficiently produced by bacterial fermentation at reasonable production costs. Recent advances in the production of short-chain-length (SCL) PHAs by bacterial fermentation are reviewed. Current status of the production of medium-chain-length (MCL) PHAs and SCL-MCL-PHA copolymers is also reviewed.

Introduction

Polyhydroxyalkanoates (PHAs, Fig. 1) are thermoplastic or elastomeric polyester with biodegradable and biocompatible properties, and can be produced from various carbon substrates by numerous microorganisms[1-7]. These microbial PHAs have recently attracted much industrial attention to be used in a wide range of consumer products, agricultural, marine, and medical applications[3,4,8]. The physical and mechanical properties of PHA can vary with the number of main chain carbon atoms and types of alkyl-pendent groups (R groups) in the monomer units[8,9]. PHAs can be divided into two groups by the number of carbon atoms in the monomer: short-chain-length (SCL) PHA consisting of 3 – 5 carbon atoms and medium-chain-length (MCL) PHA consisting 6 –14 carbon atoms[9,10].

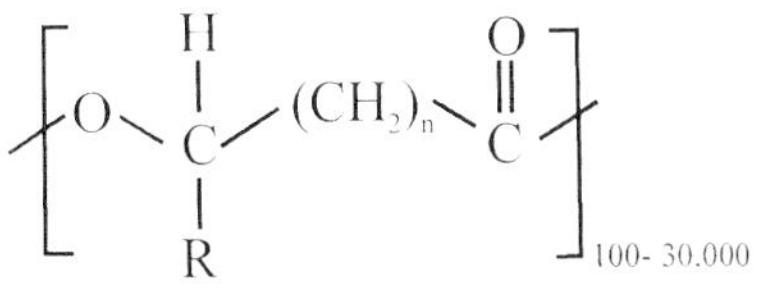

$$\left[O \diagdown \underset{\underset{R}{|}}{\overset{\overset{H}{|}}{C}} \diagup (CH_2)_n \diagdown \overset{\overset{O}{\|}}{C} \right]_{100-30,000}$$

Fig. 1: General structure of polyhydroxyalkanoate.

        CCC 1022-1360/00/$ 17.50+.50/0

Although many different PHAs possessing useful properties have been discovered, they are currently too expensive to produce[4,11]. Commercial uses of PHAs have so far been limited to specialty applications where properties are more important than costs. Much current research focuses on the development of strategies for the efficient production of different types of PHAs[4,7]. In this paper, recent advances in the production of PHAs with high productivity and content are presented. Particularly, it was aimed to review the current status of producing SCL-PHAs, MCL-PHAs, and SCL-MCL-PHA copolymers.

Poly(3-hydroxybutyrate)

Poly(3-hydroxybutyrate) (PHB) is the best characterized member of PHAs. PHB is often compared with isotactic polypropylene because it is a enantiomerically pure polymer with a single methyl substituent adjacent to the methylene group placed regularly along its backbone[8]. PHB isolated from bacterial cells usually possesses very high number-average molecular weight (approximately $10^5 - 10^6$) and high crystallinity (more than 50%)[8,12]. A wide variety of prokaryotic organisms accumulate PHB when growth is limited by the depletion of an essential nutritional component such as nitrogen, oxygen, phosphorous, sulfur, or magnesium in the presence of excess carbon source. However, only several bacteria such as *Ralstonia eutropha*, *Alcaligenes latus*, *Azotobacter vinelandii*, methylotrophs, and recombinant *Escherichia coli* can be employed for the efficient production of PHB mainly because high PHB content and high PHB concentration can be obtained with these bacteria[3,10].

A. latus grows fast, accumulates PHB during growth, and utilizes sucrose, and thus inexpensive raw sugar and beet/cane molasses[10,13]. However, the PHB content that could be obtained was typically less than 50% of cell dry weight[13]. To increase the PHB content, the strategy of applying nutrient limitation was examined during the cultivation of *A. latus*[14]. After nitrogen limitation was applied, the increase of cell concentration was only due to the increase of PHB concentration. By the fed-batch culture with nitrogen limitation strategy, PHB content could be increased to 88 wt% of dry cell weight. With this strategy, the PHB productivity of 4.94 g/L-h, which is the highest value reported to date, was obtained[14].

Recombinant *E. coli* has been investigated for the production of PHB because it has several advantages as reviewed elsewhere[15,16]. By the fed-batch culture of recombinant *E. coli* harboring the *R. eutropha* PHA biosynthesis genes and *E. coli* *ftsZ* gene, cell dry weight of 206 g/L, PHB concentration of 149.7 g/L, and PHB content of 73% were achieved in a chemically defined medium, resulting in the PHB productivity of 3.4 g/L-h[17]. Recently, we constructed recombinant *E. coli* strains harboring the newly cloned PHA biosynthesis genes from *A. latus* and examined production of PHA by the recombinant *E. coli*[18]. By the fed-batch culture of recombinant *E. coli* harboring the *A. latus* PHA biosynthesis genes, the final cell and PHB concentrations of 194.1 g/L and 141.6 g/L, respectively, were obtained in 30.6 h, resulting in a much higher productivity of 4.63 g PHB/L-h[18].

From the economical evaluation of the processes for the production of PHB by various bacteria, it was found that these processes for the production of PHB by *A. latus* or by recombinant *E. coli* harboring the *A. latus* PHA biosynthesis genes were more economical compared with the processes by other bacteria[19,20].

Poly(3-hydroxybutyrate-co-3-hydroxyvalerate)

PHB is a highly crystalline material mainly due to its stereoregularity, and its 3-substituted structure makes it thermally unstable at temperature immediately above the melting point[8]. Both of these characteristics suggested potential difficulties of employing PHB in plastics and fibre application. The breakthrough came through the development of a family of copolyesters based on 3-hydroxybutyrate (3HB) and 3-hydroxyvalerate, referred to as poly(3-hydroxybutyrate-co-3-hydroxyvalerate) [P(HB/V)][21]. P(HB/V) had been produced commercially by fed-batch culture of *R. eutropha* from glucose and propionic acid by Imperial Chemical Industries under the trade name Biopol[21].

There have been a series of papers on the production of P(HB/V) by the wild type producers[10]. Recently, a high level production of P(HB/V) by the fed-batch culture of recombinant *E. coli* was also reported[22]. For the production of P(HB/V) copolymer by recombinant *E. coli*, co-substrate such as propionic acid was also required to provide 3-hydroxyvalerate precursors[23]. Acetic acid and/or oleic acid induction resulted in the enhanced production of P(HB/V) in both flask and fed-batch cultures[24]. By the fed-batch culture of recombinant *E. coli*, P(HB/V) concentration, P(HB/V) content and P(HB/V) fraction of 158.8 g/L, 78.2 wt%, and 10.6 mol%, respectively, could be obtained in 55.1 h, resulting in a high P(HB/V) productivity of 2.88 g/L-h[22]. These results together with the simple purification method recently developed for recombinant *E. coli*[20,25] demonstrated the possibility of producing P(HB/V) at a cost much lower than previously thought. Also, the P(HB/V) fraction could be varied by varying the propionic acid concentration in the feeding solution during the culture[22].

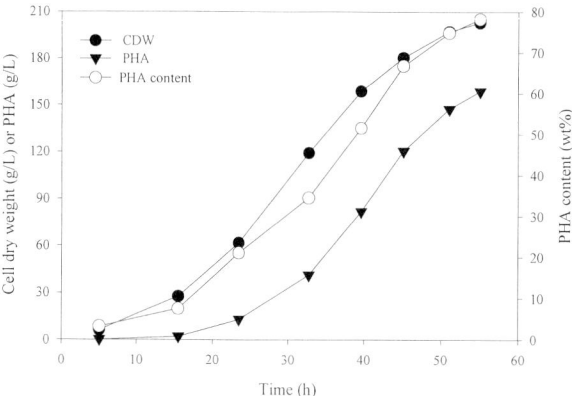

Fig. 2: Time profiles of cell dry weight, PHA concentration and PHA content (wt%) during the fed-batch culture of XL1-Blue (pJC4) with oleic acid supplementation after acetic acid induction. The feeding solution was added to increase the concentrations of glucose and propionic acid to 20 g/L and to 5 mM, respectively, after each feeding (reprinted from ref. 22 with the permission).

It has also been reported that several other SCL-PHA copolymers such as poly(3-

hydroxybutyrate-co-4-hydroxybutyrate) could be produced, even though at much lower efficiency[7]. Because these copolymers possess superior polymer properties to PHB homopolymer, development of efficient fermentation process for the production of these copolymers will allow development of a wide range of applications.

Medium chain length-PHAs

Medium chain length (MCL-) PHAs are rubbery and flexible material with low crystallinity, and can be used in a wide range of applications which cannot be fulfilled by PHB and other SCL-PHAs[26]. *Pseudomonas* strains have been employed for the production of MCL-PHAs from a number of organic compounds such as alkanes, alkenes or alkanoic acids[3,10]. However, to date, the final MCL-PHA concentration and PHA content obtained have been at relatively low levels compared with those of SCL-PHAs, which hampered the development of their applications[27,28]. By the fed-batch culture of *Pseudomonas oleovorans*, high cell concentration of 112 g/L was obtained, but the PHA content was less than 25 wt%[27]. Recently, efficient production of MCL-PHA by *Pseudomonas putida* has been reported[29]. By the application of phosphorus limitation and optimized feeding strategy, PHA concentration and PHA content obtained in 38 h were 72.6 g/L and 51.4 wt%, respectively, resulting in a high PHA productivity of 1.91 g/L-h[29]. The molar fractions of 3-hydroxyhexanoate (3HHx), 3-hydroxyoctanoate, 3-hydroxydecanoate, 3-hydroxydodecanoate, and 3-hydroxy-5-cis-tetradecanoate in PHA were 10.1 mol%, 37 mol%, 34.7 mol%, 8.6 mol%, and 9.6 mol%, respectively[29].

Several bacteria produce PHAs consisting of both SCL- and MCL-hydroxyalkanoate monomer units[30-32]. Some *Pseudomonas* strains accumulate PHAs consisted of 3-hydroxyalkanoates (3HAs) of C_4 to C_{12}[31], and *Aeromonas caviae* produces a copolymer composed of 3HB and 3HHx[32]. These copolymers were found to possess useful material properties of their own[33]. *Aeromonas hydrophila* isolated from raw sewage samples was found to produce a copolymer composed of 3-hydroxybutyrate and 3-hydroxyhexanoate[34]. By the fed-batch culture of *A. hydrophila*, PHA concentration and PHA content obtained in 43 h were 43.3 g/L and 45.2 wt%, respectively, resulting in a high PHA productivity of 1.01 g PHA/L-h[34]. The fraction of 3-hydroxyhexanoate in PHA was 32.1 mol% at maximum.

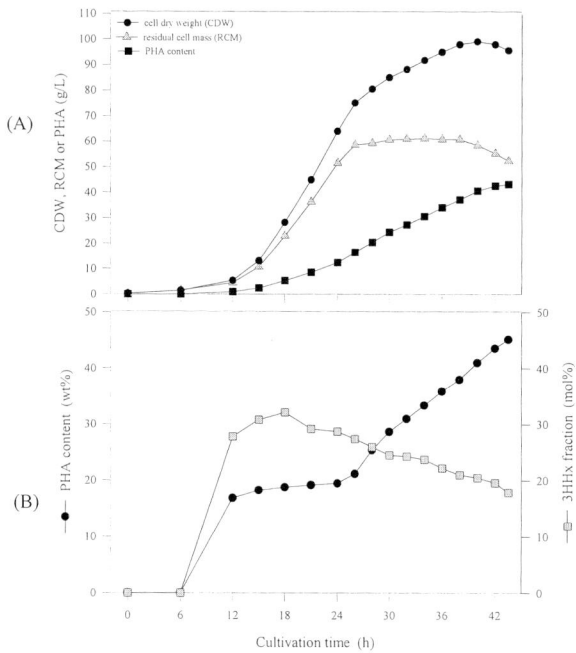

Fig. 3: Time profiles of (A) cell dry weight, residual cell mass, and PHA concentration and (B) PHA content, and 3HHx fraction during fed-batch cultivation of *A. hydrophila* (reprinted from ref. 34 with the permission).

Conlcusion

PHAs have been thought to be good candidates for completely degradable plastic and elastomeric materials. However, their high production cost has limited their development. To reduce the high production cost, much effort has been devoted to the isolation of better bacterial strains and development of efficient fermentation and recovery processes. PHB and P(HB/V) copolymer can now be produced efficiently with the production cost less than $ 3.5 per kg. Reports on the production of other SCL PHAs, MCL PHAs, and SCL-MCL PHAs to high concentrations with high productivities are beginning to appear. With further development, various members of PHAs will be produced at reasonable costs, which will allow PHAs to become a leading biodegradable polymer.

Acknowledgement

The authors' work has been supported by the Ministry of Science and Technology and by the LG Chemicals, Ltd.

References

1. A.J. Anderson, E.A. Dawes, 1990. *Microbiol. Rev.* **54,** 450 (1990)
2. Y. Doi, *Microbial polyesters*, VCH, New York, p.6 (1990).
3. S.Y. Lee, *Biotechnol. Bioeng.* **49,** 1 (1996)
4. S.Y. Lee, *Trends Biotechnol.* **14,** 431 (1996)
5. S.Y. Lee, J. Choi, in: *Manual of Industrial Microorganisms and Biotechnology*, 2nd ed., American Society for Microbiology. A. L. Demain and J. E. Davies (Eds.) p. 616 (1999).
6. S.Y. Lee, J. Choi, in: *Metabolic Engineering*, S. Y. Lee and E. T. Papoutsakis (Eds.) Marcel Dekker, p. 148 (1999).
7. A. Steinbüchel, B. Füchtenbusch, *Trends Biotechnol.* **16,** 419 (1998)
8. P. A. Holmes, in: *Developments in crystalline polymers*, Vol. 2. D. C. Bassett (Ed.), Elsevier, London. p. 1 (1988)
9. A. Steinbuchel, H.E. Valentin, *FEMS Microbiol. Lett.* **128,** 219 (1995)
10. S.Y. Lee, H.N. Chang, *Adv. Biochem. Eng. Biotechnol.* **52,** 27 (1995)
11. J. Choi, S.Y. Lee, *Bioprocess. Eng.* **17**: 335 (1997)
12. S.K. Hahn, Y.K. Chang, S.Y. Lee, *Appl. Environ. Microbiol.*, **61,** 34 (1995)
13. T. Yamane, M. Fukunage, Y.W. Lee, *Biotechnol. Bioeng.* **50,** 197 (1996).
14. F. Wang, S.Y. Lee, *Appl. Environ. Microbiol.* **63,** 3703 (1997).
15. S. Fidler, D. Dennis, *FEMS Microbiol. Rev.* **103**: 231 (1992)
16. S.Y. Lee, *Nature Biotechnol.* **15**: 17 (1997)
17. F. Wang, S.Y. Lee, *Appl. Environ. Microbiol.* **63,** 4765 (1997)
18. J. Choi, S.Y. Lee, K. Han, *Appl. Environ. Microbiol.*, **64,** 4897 (1998)
19. S.Y. Lee, J. Choi, *Polymer Degrad. Stabil.* **59,** 387 (1998)
20. J. Choi, S.Y. Lee, *Biotechnol. Bioeng.*, **62,** 546 (1999)
21. D. Byrom, *Trends Biotechnol.* **5,** 246 (1987)
22. J. Choi, S.Y. Lee, *App. Environ. Microbiol.* **65**: 4363 (1999)

23. K.S. Yim, S.Y. Lee, H.N. Chang, *Biotechnol. Bioeng.* **49,** 495 (1996)

24. K.S. Yim, S.Y. Lee, H.N. Chang, *Korean J. Chem. Eng.*, **12,** 264 (1995)

25. S.Y. Lee, *J. Microbiol. Biotechnol.*, **6,** 147 (1996)

26. K.D. Gagnon, R.W. Lenz, R.J. Farris, R.C. Fuller , *Rubber Chem. Technol.* **65,** 761 (1992)

27. M.B. Kellerhals, W. Hazenberg, B. Witholt, *Enzyme Microb. Technol.* **24,** 111 (1999)

28. R.A. Weusthuis, G.N.M. Huijberts, G. Eggink, in: *1996 International symposium on bacterial Polyhydroxyalkanoates*, G. Eggink et al, (Eds.) Davos. NRC Research Press, Canada. p 102 (1996)

29. S.Y. Lee, H.H. Wong, J. Choi, S.H. Lee, S.C. Lee, C.S. Han. submitted.

30. H. Brandl, E.J. Knee, R.C. Fuller, R.A. Gross, R.W. Lenz. *Int. J. Biol. Macromol.* **11,** 49 (1989)

31. M. Kato, H.J. Bao, C.K. Kang, T. Fukui, Y. Doi, *Appl. Microbiol. Biotechnol.* **45,** 363 (1996)

32. T. Fukui, Y. Doi, *J. Bacteriol.* **179,** 4821 (1997)

33. Y. Doi, S. Kitamura, H. Abe, *Macromolecules* **28,** 4822 (1995)

34. S.H. Lee, D.H. Oh, W.S. Ahn, Y. Lee, J. Choi, S.Y. Lee. *Biotechnol. Bioeng.* in press (1999).